T0304884

Rough Patch

How a year in the garden brought
me back to life

Rough Patch

How a year in the garden brought
me back to life

KATHY SLACK

ROBINSON

ROBINSON

First published in Great Britain in 2025 by Robinson

1 3 5 7 9 10 8 6 4 2

Copyright © Kathy Slack, 2025
Illustrations by Rosie Ramsden

The moral right of the author has been asserted.

All rights reserved.
No part of this publication may be reproduced, stored in a retrieval system,
or transmitted, in any form, or by any means, without the prior permission
in writing of the publisher, nor be otherwise circulated in any form of binding
or cover other than that in which it is published and without a similar condition
including this condition being imposed on the subsequent purchaser.

A CIP catalogue record for this book
is available from the British Library.

ISBN 978-1-47214-885-8

Designed by Clare Sivell
Typeset in Minion Pro by Clare Sivell

Printed and bound in Great Britain by Clays Ltd, Elcograf S.p.A.

Papers used by Robinson are from well-managed forests
and other responsible sources.

The authorised representative	Robinson
in the EEA is	An imprint of
Hachette Ireland	Little, Brown Book Group
8 Castlecourt Centre, Dublin 15,	Carmelite House
D15 XTP3, Ireland	50 Victoria Embankment
(email: info@hbgi.ie)	London EC4Y 0DZ

An Hachette UK Company
www.hachette.co.uk

www.littlebrown.co.uk

For Hadleigh.
And Humf.

Contents

A clutch of beige seeds
transformed into riotous life

Foreword

Years ago, in a bid for freedom, I left my small flat in London for the open air of the Cotswolds. Amid the clutter of moving boxes, I ventured out into the garden of our new cottage in West Oxfordshire. This was what we came for. Green space. I had yearned for it, had realized the hanging baskets that clung to the stairwell of our basement flat would never suffice, had decided it was worth the commute to my London office, and had taken the leap.

I cleared a little square of ground in our new garden, marked it out with tent pegs and string, and allocated it to vegetables to tide me over until we'd installed raised beds. In those early months, my pockets were often filled with seeds that would later be liberally scattered over the patch. At weekends I would thumb my dog-eared RHS manual, kneeling on the soil and wondering if 'plant to a depth of 5cm' meant from the bottom of the centimetre-long broad bean seed or the top. I liked to be precise. Despite my haphazard care and wandering attention, the seeds grew. Over a few seasons, and with much trial and error, I became a competent novice vegetable grower, and would take gluts of beans and courgettes into work to foist on anyone who didn't look at me like I was mad. But the people I worked with lived off takeaways and stored shoes in their ovens; they must have thought I was from another planet.

Later, when it seemed like everything in my life had fallen apart, this patch was where I returned to on the days I could manage to get off the sofa. By then, sick, off work and drugged-up, I found that, on a good day, I could sit among the vegetables, tugging on the odd weed, planting the occasional seed. The draw of nature that had motivated me to move out of London all those years ago was now pulling me towards the veg patch, again in search of respite. And I found it. I put my hands in the soil and, for a second, there was silence. No grief, no panic. Just stillness.

Ultimately, this small plot of land showed me a new way to exist in the world. A new reality. Though I initially thought the vegetable patch was my escape from reality, I soon realized I was wrong. The veg patch *was* reality. I was not 'getting away from it all' when I came to the patch; I was putting myself back in, immersing myself at the centre of things. And, very slowly, over many months, my sense of peace grew as the plants grew. I didn't know then that this was the beginning of recovery and the start of a whole new life. That over the course of a year, the veg patch, very quietly, without fanfare, would save my life. It would reconnect me with the world, show me a new path and give me hope – a battery hen let loose and shown the free-range world.

Scribbled down here are the many lessons the veg patch taught me. It is broken up month by month following the ebb and flow of the Cotswold seasons, beginning in June (its perkiness in stark contrast to my mood). The story that follows is my recollection of that year and the balm that my little patch of earth provided. And it all started with a few of those lifeless little seeds in the palm of my hand.

Introduction

It is incredible, nature, when you stop to think. Even at its most ordinary. Take growing vegetables, for example. One morning in March, or thereabouts, you take a handful of unassuming, apparently lifeless seeds and scatter them over the ground in your vegetable patch. By July, if not before, this unpromising handful has turned into a jungle. A *jungle*. There are gargantuan triffids galumphing about your allotment, tangling themselves up with one another, drooping under the weight of their harvest. A clutch of beige seeds transformed into riotous life. And that's when it really hits you: you are witnessing magic. Proper Jack-and-the-Beanstalk alchemy. It is wondrous to behold. And even though I have been growing food for eleven years I marvel at it every time.

To start with a seed and finish up with supper is spellbinding. I have found it restorative too. Therapeutic, even. I think that growing and eating your own food, even if that's only a pot of herbs on a windowsill, is as good for the mind as it is for the body.

Don't get me wrong. I'm not about to claim that kale cures everything. No 'ditch the Prozac and eat blueberries instead' mumbo jumbo here. Not on my watch. No, I mean that growing vegetables is, for me, a powerful antidote to the pressures of life. A way to re-wild my mind, reconnect with nature and lift myself out of the barrel of treacle that is life.

You know, Life. In my experience it can get a bit much. Eleven years ago I was a Global Strategy Director in a London advertising agency. I had been promoted quickly and I was a fully-fledged jet-setting, Jimmy Choo-wearing, BlackBerry-addicted achiever. I had an exciting, well-paid job, a family who loved me, and a beautiful, freshly renovated home in the Cotswolds. And then it all went horribly wrong.

I had gone into advertising on a graduate recruitment scheme aged twenty-three-ish, and had flitted between agencies every couple of years, working my way up the ladder each time I moved. My job, which inexplicably was called being a Planner – why I don't know, we never planned anything – involved helping big companies work out what to say to consumers to make them buy a product, and then creating a thrilling and single-minded brief for creative teams in the agency to write adverts that communicated whatever had been decided upon.

A good planner is able to see through the undergrowth of

data and research to a simple, perceptive insight about the product or the consumer. You then express this insight with a pithy phrase, thus unlocking boundless creativity and providing the foundation for amazing work. For Apple in the 1980s, that insight was 'we don't make beige boxes (like IBM do)', we make 'tools for creative minds' and it was the springboard for the famous 'Think Different' campaign. Many clients got promoted off the back of that insight. A lot of creatives won a lot of awards. Apple's sales soared. This is the power of the Planner. And so it is inevitable that some planners have an aura about them, a reputation for being bookish and intellectual, for being slightly removed from day-to-day logistics so that they can see the big picture, the big idea. It is OK for this sort of planner to be late to meetings, or not know how to use a conference phone, or not book their own travel; they are probably thinking great thoughts and had best not be disturbed. Things have changed now, but when I was in Ad-land, this sort of planner had the door to their ivory tower held open for them and were asked, as they glided up the stairs, if they would like tea brought up.

I was not this sort of planner. I was the other sort of planner. The sort who works on more conventional, but often bigger, clients and whose job is more chief-negotiator than big-thinker. Here, the task is to sew together all the disparate priorities and contradictory requests from various parties in different corners of the corporate behemoth to create something resembling one idea that keeps everyone happy and has a hope in hell of setting the creative team off towards a plausible concept.

These are the bread-and-butter accounts that pay salaries in the agency but don't win the awards. Certain I wasn't learned enough to get a seat in the ivory tower with the cool kids, I'd worked my way up by leaning into accounts like this, specializing in big brands within FMCG (fast-moving consumer goods – stuff you buy in the supermarket). Most of these brands are owned by one of a handful of multinationals who make everything from margarine to washing detergent. Each has a particular process for deciding on marketing messages or new product launches and those processes are fiddly and obtuse. You can understand why: the reasons people buy a particular soap in India are totally different from the reasons they buy it in Australia, but you still need to find one product, one pack design and, ideally, one advertising campaign to suit them all. However smart the client, however successful the brand, it is always, *always*, a bugger's muddle.

The image of Ad-land is a world of wild parties, long lunches, outrageous expense accounts and hard-talking creative types in edgy-but-expensive suits. And sure, there's lots of travel, business-class flights, fancy hotels and some nice outfits. I went to Petra in Jordan and the Great Wall of China with work. I have eaten in the best restaurants in Stockholm, visited the coolest underground coffee houses in Poland, jogged along a beach in Puerto Rico at dawn, all thanks to work trips. It was one hell of a ride. Plus, you get paid really well and have a good, predictable career progression. It would be the height of middle-class privilege to call it a bad job (and by the way, my diamond shoes are pinching a little).

But. It is also slow and mundane and frustrating, like most jobs. I worked at one agency for two years and never made an advert. Not a single TV spot. We wrote scripts, researched ideas, tested mock-ups, but the client's decision-making process was so bureaucratic that nothing ever got signed off. In fact, we shot a whole campaign, including a very, *very* expensive TV ad, and it never aired. *Mad Men* it is not.

It was cut-throat too. People came and went in a heartbeat. There were no written warnings, no 'points for development' in your annual review, no solemn one-to-ones with your line manager (I don't think I ever knew if I even *had* a line manager). You just got fired. At one agency it even had a name: *the white van.*

'Where's Bob?' someone would ask. (I've made the name up; people in Ad-land are too cool to be called Bob.)

'Oh, the white van came for him.' Meaning he had been fired and paid off, so no one ever saw or heard from him again. Poor Bob.

Another time, at the same agency, I stood at a staff party, nervously sipping a drink, while a senior Planner talked at me. He was high up in the management team of the agency, in charge of hiring and firing, and he was advising me about the trials and tribulations of Getting Ahead. 'You know, every Planner I have called into my office to promote has afterwards confessed they thought I was about to fire them.' He was gleeful, delighted by his power to terrify. I laughed obediently while my moral compass cast about looking for north. What kind of culture makes amazing people who are about to be promoted feel so inept they think they're getting fired?

Anyway, for better or worse this was my world for over a decade. I relished the challenge, especially of those unwieldy multinational FMCG accounts. I thought I had a talent for inhaling all the complexity, processing it singlehandedly and then spitting it back out into a simple and inspiring format that the creative teams could turn into entertaining adverts. I knew the Pret breakfast menu by heart and I could format a kick-ass PowerPoint presentation. They were beautiful. Real works of art. I was winning.

Pret Ham and Cheese Croissant

There was very little I missed about London when I finally left, but this was one of them. The Pret local to my office opened at 7 a.m. and I would be waiting outside as they pulled the shutters up to get the hottest, freshest ham and cheese croissant to eat at my desk. Stodge, warmth, carbs, salt – everything an ill-slept body needs.

I recreated them with some success at home by slicing a 320g sheet of ready-rolled puff pastry into four rectangles, spreading the middle of each with a tablespoon of tomato chutney and topping with a square of ham and some grated cheddar. I folded the opposite corners over the filling, brushed the pastry with beaten egg and baked them in the oven at 195°C for 15 minutes.

Eaten hot from the baking tray or left to cool, frozen then reheated as needed, they tasted even better than the original. But perhaps that was because I wasn't eating them at my desk anymore.

So, there I am with my fancy heels, my corner office, my excellent PowerPoint skills and I am climbing the ladder to join the board of my most recent agency, where I have been for almost a year. My husband, Paul, and I had moved out of London to a cottage in the Oxfordshire Cotswolds a few years into our careers and now did the City commute.

I couldn't get away from London fast enough when we moved. I was desperate to leave. I felt an ache for the greenery, the fields, the space, the skies of the country. It was visceral. Which is daft really because I'd always lived in small-town middle England and I wasn't especially outdoorsy. I had no experience of rural life. But, somehow, I just knew I needed to surround myself with nature. Which should have set alarm bells ringing then. Perhaps someone who feels the pull of countryside life so strongly might not naturally flourish in concrete Ad-land? Had I questioned why I so urgently needed to leave, I would have realized that moving to the countryside was, in fact, fleeing from the stress of the city. Had I examined my thoughts more, I would have seen that I was desperately searching for some respite and comfort that would make my unmanageable life more manageable. Then, as now, the first thing I look for when I need rest and retreat is nature.

The upshot of this move was that, in the week, I commuted to town every day and the weekends felt like rehab. Back then, it was a four-hour round trip. But that's no problem, right? I could work some of my fourteen-hour day on the train, and dialling in to conference calls remotely is easy (though it does not endear you to fellow passengers). And the walk from the office to the station along the canal is quite pretty at sunset. And anyway, I'm on a plane most of the time and the drive from West Oxfordshire to Heathrow is super-quick at 5 a.m. so it's all fine, really. Really. Fine.

The burn out was so predictable, in hindsight. I had been stressed for a decade and jet-lagged solidly for a year. Like so many others, over the course of a few weeks my body and my brain shut down. Gradually but inexorably, my mind unravelled. There was no dramatic scene. No public meltdown that made it obvious I wasn't coping. Instead, there was a slow, insidious creep of despair, terror and disfunction until I no longer recognized myself.

Partly it was cognitive. I began to forget things – words, meetings, how I'd got to work, what day it was. As the fog descended, I couldn't compute anything properly – how to send an email, how to start the car. It was like when you haven't done something for a while – ride a bike, speak French – and the knowledge of how to do it is now just out of reach.

But the dismantling was also emotional. I felt profoundly sad. Without reason or focus, I was consumed by grief, distraught at how inexplicably alone I felt. I would dwell for hours on what a failure I was, how worthless and unlovable I had become. I couldn't have given you a reason or a cause for these feelings. They were like phantom emotions, but they were the strongest emotions I have ever felt in my life, to this day.

It was physical too. In what I now know to be a common response to stress, my body felt constantly alert, ready for flight. I was edgy, fizzing with the jitters; every sense was heightened; things were brighter, louder, faster, more urgent, more dangerous, more intense, and all of it was overwhelming. If that ever abated, I became the polar opposite: lethargic and exhausted, a lead weight in my chest and glue in my joints so that every move was a claggy effort.

After a routine trip to the nurse escalated – more of which later – I was diagnosed with depression and signed off work. One week I was flying to Dubai for a lunch meeting with the board of a multibillion-dollar company and the next I was sitting on my sofa unable even to make a cup of tea and planning my suicide. I would remain on that sofa, tea-less, for almost a year. And rehabilitation would come from an unlikely source.

What follows is the story of how it all happened. But that's not really the point. The point is to try and understand why growing food fixed me. Why not walking? Or getting a dog (though that *was* a big deal)? Or literature? Or just any old gardening? What was it specifically about growing vegetables that made such a difference to my well-being in a way that these other things did not? Why does this way of connecting with nature, even now, offer a very specific kind of solace in difficult times?

I am interested to work this out, partly for my own self-indulgent curiosity, but also because I want to know what to do if it happens again. If the allotment was a coping mechanism I want to know what made it such a successful one, so that if I am ever again 'walking around with your head on fire and no one can see the flames', as Matt Haig puts it, I know what a good fire blanket looks like.

It is helpful too, I think, to collect the lessons I learnt in the veg patch. Because apart from the solace it offered, growing food has also taught me about success, self-reliance, compassion, kindness and more. Nature has become the foundation of my core values and here is an opportunity to reflect on what I've learnt.

I have also found there is some comfort in knowing you are not the only one who has felt like this, that others have been where you are and have made it out of the pit. Which is why I think it's good to talk about mental health, and especially to share ways of coping with it. That's not to say what worked for me will work for everyone; melancholia is notoriously individual and what I found calming, like weeding, might drive others to distraction. But even if your cure isn't my cure (if indeed it can be cured at all), then it's useful to know that finding an appropriate remedy feels like coming home rather than escaping. And, if nothing else, it's a comfort to know we are not alone.

The recent willingness to talk about mental health is, without question, A Good Thing. But it comes with its dangers.

The first would be voiced by my Yorkshire grandfather, were he still alive.

'What do you want to go picking that scab for? Leave well alone and it won't scar.'

Well, Grandpa, your caution (let's call it that rather than cynicism) is well-founded. For if we spend too much energy examining our own anguish, might we not make it worse? Hadn't we better leave well alone? What if we talk about it so much that we define ourselves by our depression? Stephen Fry described this in an interview with Ferne Cotton brilliantly: 'I still feel it occasionally, a danger of becoming sort of professionally mentally unstable, that that's what I am, that's who I am.'

Wise words both. So, Grandpa, I will not wallow.

My family's second favourite, and deeply Yorkshire motto is: 'Don't go puffing yourself up.' Meaning, don't go talking about yourself too much because it makes you look like you think you're special. In other words, talking about your feelings is a bit self-indulgent and vain. Which is daft really because my family are all a bunch of softies who will cry at the very mention of a Richard Curtis movie. Still, I do agree that picking over the entrails of a messy breakdown could be narcissistic, were that the end game. Why should anybody else care? But I am going to remove my metaphorical family crest – puffing-up motto there inscribed – from the wall and put it in a drawer for a bit, because that is *not* the end game. The purpose here is *hope*. Hope offered by nature (always present, always to be relied upon, always free) to bring peace and perspective in times of pain and to remind us that the sun will still rise. And that's worth risking a bit of hereditary eye-rolling.

Plus, the non-specialness of my situation is kind of the point. This isn't a particularly dramatic tale. Nothing extreme or drastic happened. I didn't have a meltdown in a client meeting or end up homeless. It was just me, sobbing on the bathroom floor. I was just a middle-class woman with a nice life. Pretty ordinary. Because that's what depression is. It's not exciting or dramatic. It's dulling, monotonous, unremitting. It doesn't have to be some huge, key plot-point-blowout moment. That I am nothing special shows how depression is everywhere and can happen to anyone in all sorts of ways. There is no barometer of shitness. Just because mine didn't involve anything you could turn into a film does not make it less intense. These are the strongest, most destructive feelings I think are possible in a human brain, and they all happened under the most ordinary and everyday of circumstances.

The other risk is that by talking openly about depression we make it seem like everyone has depression; that it is quite normal and not a big deal. When, in fact, it is life-threatening.

Or it becomes some macabre woke fashion-accessory, the badge of a sensitive, interesting soul. Like a gap year in India. Packaging depression up tidily to make it sellable and easily consumed suits the wellness industry because a neat problem can have a neat solution: retreats, supplements, mindfulness apps, books, diets, de-cluttering programmes, life coaching. Do this and you'll be cured!

Take it too far and depression risks becoming nothing more than a predictable narrative arc in someone's life, an understandable moment of suffering that turned out to be the making of them. But depression is none of those things. It might be widespread, but it is not ordinary. It is not an unpleasant phase we all go through, like chicken pox. It isn't linear or orderly or predictable. It is a mess. An ugly, unpredictable, unedifying train wreck you would not wish upon your worst enemy.

And what if we start to over-diagnose and imagine that *any* unhappiness is depression? Again, cash registers ring for the wellness industry: a bigger audience to sell to. Sometimes feeling depressed is a perfectly logical response to a dreadful situation. But it is not clinical. Hyper-sensitivity to melancholy is a particular risk for those who look after anyone who has been depressed before. As soon as they look even the tiniest bit glum, the worry starts. 'Is this sadness a reasonable reaction to something rubbish or are they depressed again?' The depressed person can think like this too. If you dwell on it for too long, you forget which way is up.

With so many pitfalls, I am tempted to stop here. But that

would not do. Not least because there is food to discuss, but also because nature has been such a powerful cure for me that I can't believe it doesn't hold potential for others too. It has helped me find my place in the world, giving me a new perspective, a new sense of conviction. With nature around me, I feel I can always tell which way is up. I am now far less likely to lose my sense of what really matters in life and far more compassionate to myself and others. It has become the underpinning of all my values, my sense of self-worth and my peace.

All that from a year of haphazardly sowing seeds in empty yogurt pots and tending wonky carrots in a makeshift veg patch. This book is a chance to reflect on what happened that year, to consider what lessons the veg patch taught me and how nature offers a unique remedy.

June

light dew clings
to the fronds of
the wild fennel

How I had everything, and it all went wrong one summer's day

IN THE VEGETABLE PATCH...

There is nothing quite like a kitchen garden in June. Unlike late summer, when the ground is cracked from drought and the plants a little yellowed from overwork, early summer is all perky neatness. The soil is dark and chocolatey rich from the summer rain and every crop is still in orderly rows having only just been planted out after the last late-May frost, not yet ruffled into disorderliness by weeds or poor husbandry. Plants dash skyward to the endless sunshine with youthful vigour, and everything is luscious, clean and green. So much green.

Who knew there were so many shades of green? The creamy green broad beans provide a backdrop, straight stemmed and strong despite the weight of their spongy pods. In front of them, curly pompoms of bottle-green kale plants which have just been transplanted from their seed trays. And in between both, the

17

livid cartoon green of courgette leaves, green beans, pumpkins, spinach, all growing so fast you can practically watch it. And amidst it all, little explosions of colour – yellow tomato flowers, mauve bean flowers and white pea flowers – which look like Christmas baubles against this greenery.

It is a busy, worker-bee time in the veg patch, when every-thing, gardener and plants, is putting its best foot forward and making an effort, hopeful of a good harvest.

Daybreak. A glorious June morning in the Cotswolds. One of those golden, hazy mornings you find in novels. In the gar-den, a light dew clings to the fronds of the wild fennel in the raised vegetable beds. The dawn chorus is starting up. House martins chirrup as they dart about over the poppy-flecked fields beyond, as if still thrilled to be back in the summer choir. Our cottage is radiant in the morning light. The pale Cotswold stone turned amber, warm and welcoming.

Inside, sleeping, is my husband, Paul. He loves me. I love him. We have families who love us both. We have enough money not to think about money. We have good health and successful careers. We are very lucky. We have everything. And everything is picture perfect.

I am sitting, pyjama-ed, on the edge of a raised vegetable bed in my garden. Among all this beauty. All this good for-tune. Sobbing. I mean, uncontrollably. I am half worried I will wake the neighbours or frighten the sheep in the nearby field. Proper bawling.

But I have no idea why. Jet lag? The day before yesterday I had flown to New York from London for a half-hour meeting with the Creative Director of our US office. In the end it was barely a fifteen-minute meeting, and it mostly involved him shouting at me. I had taken a daytime flight, had the meeting/kicking, gone straight back to JFK, taken the red-eye home, showered at the airport and made it to a 9 a.m. briefing in my London office. The kicking was a bit unexpected, but the schedule was nothing new.

Or maybe it's the blinking red light from the BlackBerry on my desk that I noticed as I padded downstairs. Each blink a warning of the barrage of emails that awaits from markets in other time zones who will all want responses to their absolutely-critically-urgent-question about margarine sales before their close-of-play, which is my pre-breakfast.

Or it might be nerves about the creative briefing I have this afternoon, at which I will attempt to convince a pair of understandably jaded men that writing a radio advert about the cholesterol-lowering qualities of a cooking-fat spray is the most exciting thing they will do this year and will definitely, *definitely* win them awards. Though, to be honest, I would consider it a win if they just let me out of their office in one piece, unlike my US Creative Director.

Whatever the reason, I have noticed lately that this is becoming a morning ritual. I wake up feeling like someone has put lead weights in my lungs. I have a nameless dread that everyone I love is dead. I sit among the vegetables in the hope of a little peace, or comfort. I get in the shower and go to work.

Which is what I had better do now, because I have a Power-Point presentation about the chocolate-buying habits of Saudi women to finish on the 06:32 to Paddington before my 8 a.m. client meeting in London.

The morning does not go well. I can't work out how to turn the shower on. It is quite a complicated mixer-tap arrange-ment, but then I have done it every day since we moved here six years ago and never struggled before. And there's a similar fog about where to put the keys to turn the car on. It is not that I have forgotten, I can see the shape of the gap where the information should go, it is that the answer hovers just out of reach. Like when you can't recall a particular word, or momentarily forget what eight times four is: it is right on the tip of your tongue, you know you know it, but it persists in eluding you. Probably jet lag.

Anyway, I work out the shower and the car and I get to the station. The train that will-not-be-stopping-please-stand-back seems much louder and faster today and I cower against the railings on the platform as it whooshes past, braced for impact like a startled rabbit. The noise is so thunderous I feel like the train is driving right through me. I am terrified and shaking, covering my face and turning towards the railings in the hope they will somehow protect me. Other commuters side-glance and shuffle down the platform away from the flinching mad woman. They give me a wide berth and I get a pair of seats to myself on the train despite it being packed. (So not all bad then.) But none of this is setting me up for my 8 a.m.

I don't even remember that 8 a.m. meeting I had been so dreading, or much about work during that period. I remember the effort of getting there, the anxiety about the day, but not the days themselves. Though I am fairly sure I was not on sparkling form.

Things continue in this vein for a few months. I travel internationally most weeks. I work fourteen-hour days. I commute four hours a day. I write a lot of PowerPoint (I dream in Power-Point too). I fail to enthuse creative teams. I make dancing on the head of a pin into an art form.

What permeates these weeks most is a sense of dread. Physical, stomach-wrenching, stuff-of-nightmares terror. I have always been anxious about performing well or getting things right, but this is different. It is inexplicable. It doesn't stem from a particular worry – a brief, a work trip, a difficult conversation with a creative team. It is a directionless, panicky, pit-of-the-stomach fear that makes you recoil from every noise in case it finally signals what you have been braced for, the sky falling in and all about being lost. I wake up with it, like a lead duvet smothering me, carry it wrapped about me all day, and then fall into bed, exhausted, and sink beneath its weight into a fitful sleep.

And then the voices start. Not, thank God, the infamous

voices of psychosis, and far less dangerous, but exhausting in their own way. In my head there is a party of all my worst critics. And, boy, are they getting on famously. They discuss my top failings, listing the things I should be worried about, cataloguing all the endless decisions I will need to make today in order to function, and laughing about how woefully inept I will be at making them. Like a ghoulish cocktail party, the chatter is incessant; they talk over each other in excitement, clamouring to be heard.

'Look!' they laugh. 'She's had every possible opportunity in life handed to her on a plate and she still can't make a go of it.' 'So true. Surely, she should be richer, thinner, smarter than she is now. If only she wasn't so weak.' 'Yes, and did you notice she ate ten biscuits during that meeting because she didn't have breakfast? What must the client have thought?' 'I know! And don't even start me on what her colleagues say about her when she's not there.'

Every day is happy hour for these critics. The bar never closes.

But why? What triggered this?

By all measures, I have an easy life. One I have chosen for myself, one I wanted. No one is forcing me to do it. I made this. I have fully engaged in the creation of it all – the job, the commute, the seniority, the pressure, the lot. So why do I feel like a hostage?

A hostage in a picture-perfect existence. I know I am loved. I have no domestic responsibilities (we are, by choice, child-free) and have enough money to pay someone else to do anything we needed doing in the house. Cleaning, ironing, DIY, gardening (the flowers and the lawn were of no interest to me; I tended the veg beds and farmed out the rest) – it is all outsourced. The only thing I have to do is go to work and I am free to enjoy my perfect life the rest of the time. That is the deal.

But eventually I can't even do that. After weeks of this slow dismantling, just getting out of bed becomes an ordeal. The issue with the shower has spread to the hairdryer, how to dress, how to make tea, how to lock the back door. I can no longer really function. And my mood is so low, I can hardly bring myself to speak.

My husband is, reluctantly, driving me to the train station now, because he rightly does not trust me to drive, and putting me on the train. During these journeys he would, very gently and kindly, ask how I was feeling, or if I thought we should do something about the situation. Depending on the day, I would either blow up at him – 'How I feel is irrelevant. I've got a huge pitch for a sliced bread brand today and we absolutely have to win it.' (We didn't.) – or I simply couldn't be bothered with the effort of trying to articulate how I felt or deciding what needed to be done about it. The task seemed too gargantuan, only slightly bigger and uglier than the effort of keeping going.

Once I make it to the office, I find any work meeting so

arduous that I sneak out of the office afterwards for a nap/sob in Regent's Park.

I take to hiding as well. Proper hide-and-seek stuff. Around late morning, it would all get too much. I was consumed with worries about how badly I was performing, how many people I must be letting down. Physically, it was as if the skin had been removed from my body and all my senses heightened, so that every jolt of the lift, every phone buzzing, every loud braying colleague seemed heart-stoppingly painful. Physically painful. Like being stung. I had to get away.

So, I would find somewhere to hide. At first, I would fabricate some fictional meeting and leave the office to take refuge in the Costa round the back of Baker Street station. In advertising, people think themselves far too trendy to go to Costa, so I knew I'd be safe from discovery. Plus, it had a table in an alcove at the back which was well protected from view. But as I got sicker, the bustle of the café and the rush of the Marylebone Road, which I had to cross to get there, was too much and I sought out alternative nooks and crannies in the office in which to hide.

My favourite was a tiny meeting room in the top of the building, which was more attic cupboard than room. The eaves were low and the lift didn't go that far so the room was only accessible via a narrow staircase, which meant you were usually safe from interruption. I would creep in, take off my heels and curl up in the corner of the room. There were chairs, but they didn't feel as safe. I needed my back up against something and a vantage point from which I could see the whole room. Even now I can imagine the humiliation if someone had ever found me cowering under the window.

'Um, hi. Look, sorry to interrupt your breakdown, but we've booked this room for a pet food debrief. I didn't even know it existed but it's the only free room with a conference call pod, so … Let me help you with your shoes.'

This had been the status quo for around four months when, one morning, I was at a routine nurse's appointment. I had moved a few conference calls and managed to get her first appointment of the morning, but it would shorten my working day, making it even more hectic, so I was already wound tight. And anyway, it was eating into my preferred sobbing-in-the-vegetable-patch time.

'Your blood pressure is quite low,' said the nurse, a plump, motherly figure with kind eyes, everything you want in a nurse. 'Are you feeling OK?'

Maybe it was because it was the Sobbing Hour, or perhaps it was her warm eyes, but I dissolved into tears and garbled something incomprehensible about work being a bit stressful. She gave me a box of tissues, deposited me in the waiting room, spoke to the receptionist and marched me straight in to see the doctor.

Who was terrifying. She was the doctor you hoped you would not get allocated to when you booked an appointment. She was fierce, dismissive and abrupt. She couldn't locate a bedside manner if it hit her with a bedpan. And it was rumoured she had told one patient that his severed arm was nothing more than a flesh wound. (OK, maybe not that last bit.) Still, she looked at me, prodded me, weighed me, took a load of blood tests, asked some probing questions and then signed me off work for the foreseeable future.

Wait, what? Not work? I can't do that.

The world will end. People are depending on me. There's a meeting about snacking habits among Hungarian teenagers today that I absolutely positively have to be at. I'm about to make it onto the board. I've just bought a season ticket for the train.

'Tough,' she replied. And that was when I realized. If this stern, no-nonsense doctor was taking it all so seriously, then I probably *was* sick.

I took my doctor's note to work. I felt like that slow, chubby kid at school who comes to PE clutching a crumpled note scribbled with some poor excuse from an indulgent parent about why they must miss games, again. (Actually, I had been that child, so I knew the feeling well.) I was soft. A fraud. Others could hack it, but not me. I had given the appearance of being a competent person, but my bosses had, barely a year beforehand, hired a dud. I was so weak that the effort of trying to hide this from them had broken me.

In the face of my embarrassment and guilt and sense of

failure and apologies, my bosses were exemplary. They told me I would not be saying any of this if I had broken my neck and that, in their eyes, this situation was pretty similar and just as serious. They told me to take all the time I needed. They sensitively and delicately ignored my suggestion that I could do some work from home. They cranked the corporate wheels into motion and arranged access to any form of treatment I wanted.

'Just send us the bills and we'll keep sending the salary,' they said as they packed me back on the train, gently prising the BlackBerry from my panicked grasp.

I still marvel at how lucky I was to have this safety net. What happens when people in this situation don't have enlightened bosses with deep pockets?

I imagined I would take a couple of weeks off, maybe a month, and then get back to normal. But then I was a bit delusional. That was not how this story would go. I did not know it yet, but in fact that day marked the end of my career. Or that particular career, at least. I had no idea how long it would take to mend or what kind of life I would have once – if – I got better. And I could never have imagined that the vegetable patch would open the door to recovery.

Dad's Chicken Broth

One of the therapies that work paid for was acupuncture. Now, I am usually cynical about such things. I put acupuncture in the same bracket as homeopathy and labelled it Snake Oil. But a worried, helpful friend recommended it and at that point I was prepared to try anything. Also, I doubt my critical faculties were up to arguing the point. It was easier to acquiesce. And thank goodness I did.

I still couldn't say what it was about acupuncture that helped. Maybe it was the pain, which is brief but enough to concentrate the mind. It was like a slap in the face to all the guests at the Critics' Cocktail Party that stunned them into astonished silence. It might also have been the very kind, wise gentleman who was quietly but intently focused on nothing but making me better for an hour every week. Or perhaps it was because he fixed the qi in my system. Whatever, something made me calmer.

The acupuncture guy gave me some advice too: avoid sugar, learn meditation, that sort of thing. All of which was sound. His other tip was to make soup; specifically, chicken broth. He believed that it was so full of nutrients and goodness that it could comfort even the most addled mind.

This recipe is my dad's chicken broth recipe. It has always been legendary in our household, and I turn to it still for comfort and solace.

SERVES 4–5

FOR THE STOCK:
1 chicken carcass
2 bay leaves
A bunch of thyme
½ onion
1 celery stick

FOR THE SOUP:
1 tbsp butter
1 tbsp extra virgin olive oil
3 celery sticks, finely diced
2 carrots, finely diced
1 leek, finely sliced
2 bay leaves
2 fat garlic cloves, crushed
175g potatoes (new or old), halved
150g mushrooms, chopped
200g leftover cooked chicken, shredded
120g frozen peas
120g frozen sweetcorn
100ml double cream (optional)

When it comes to broth, stock is everything. A stock cube won't cut it, and making your own is really very simple. I know some chefs make a fuss of stock, skimming it and whatnot, but this method suits me just fine. I always make a batch with the bones left over from a whole roast chicken, freezing the resulting stock until needed. And don't worry about using cooked bones for stock. Whatever they say on *MasterChef* about raw bones being essential for stock, a cooked carcass is perfectly serviceable and why waste it? Plus, you can pick the leftover meat off the bones to add to the soup later too.

So, having banished your stock anxieties, find a pan big enough to contain the bones without them poking out over the top of the pan. Put the bones in this pan together with the other stock ingredients. Cover with cold water (about 1.5–2 litres should do it) and bring to a very gentle simmer, the surface of the liquid only just blipping away, with no lid on, for 2 hours.

Meanwhile, in another large pan or casserole pot, melt the butter and olive oil together over a medium heat, add the diced celery, carrot, leek, bay leaves and a pinch of salt and sweat, lid on, for 10–15 minutes. The vegetables should be translucent and not browned. Add the garlic in the final couple of minutes.

Once the stock is ready, strain the liquid into the casserole pot of vegetables and discard the bones, etc. Add the potatoes, mushrooms and chicken to the pot.

Bring the broth to a gentle simmer and leave to bubble for 20 minutes, or until the potatoes are cooked.

Stir in the peas and sweetcorn (they only need a minute to defrost in the broth) and check the seasoning; it usually needs more salt and a twist of pepper. (At this point you can add any greens or wiltable veg – chard, kale, etc. – you have to hand that need using up.)

For a richer broth, pour in enough cream to make the soup silky and luscious. You can add some chopped herbs, like parsley, if you are feeling fancy.

Serve in warmed bowls with hunks of bread. Comfort in a bowl.

July

something is coming

Escape?

IN THE VEGETABLE PATCH...

Sit. Just sit and marvel at the wonder of creation. That's what the July garden is for. The green beans clamour for your attention: 'Pick me now or I'll be a woody stick by tomorrow.' But the courgettes bellow even louder, huge yellow flowers flashing like amber warning signs, demanding to be seen and dealt with. The sweet peas, gossamer pastel petals filling the house with a heady, soporific fragrance, are almost oppressive; the soapy tyranny intensifies with daily picking. Eight weeks ago there was almost nothing here, and today there is life. Life everywhere and everything needs harvesting daily.

This is the month the beautiful destroyers arrive. Cabbage white butterflies. Nothing is so lovely, so perfectly English, as watching a pair of flirting butterflies with their greeny-white wings flitting and dancing above the raised beds. But the result of this picturesque waltz is a plague of green caterpillars who eat brassicas with all the vigour of a vegan Augustus Gloop, turning equally fat in the process. Picking the bloated green monsters off the kale and squishing them under a wellie is a special kind of therapy.

As if in malicious irony, the sun shone every day that July. The world was bright and green, busy and bustling, all picnic blankets, suntan lotion and lazy afternoons under a tree. In the air, the smell of grass cuttings and the thrum of the hay-balers in the fields. Every village had a fete on; bunting, scarecrows, egg-and-spoon races. It was a halcyon dream plucked right out of *Cider with Rosie*: alive and bright and carefree. And I felt exactly the opposite.

I was sad. Profoundly sad. As you would expect with depression, I suppose. Presumably it was all chemicals, the by-product of a brain misfiring, but I felt bereft and grief-stricken. I cried at everything and nothing. Even a box set of *Friends* set me off when Ross and Rachel split up. (I mean, I do cry at films a lot, but this was ridiculous.) This melancholy all seemed so silly and ungrateful when I knew everything was really fine. Intellectually, I could see all was well, but emotionally I was inconsolable with grief, a physical ache for something unknown but lost.

My memories of those feelings are clear and strong; they were so visceral, so intense that they are seared into my brain, forming a kind of scar tissue so that even now I can trace the shape of them in my head and recall how forceful and all-consuming they were. But my recollection of what actually *happened* during those first few weeks is sketchy. The days blended into one.

My scowling GP turned out to be as effective as she was terrifying and had started me on antidepressants, but they can take six weeks to kick in so were having no effect initially. There were sleeping pills too, which did at least make me unconscious at night, but I wouldn't call it sleep. They offered no rest, and I would wake groggy and hungover rather than refreshed. As well as the acupuncturist, the GP referred me to a therapist who I saw weekly. She worked at one of the private Oxford hospitals and preferred a cognitive behavioural therapy (CBT) approach. More of which later.

All this had been organized very quickly by my private health provider and was considered 'emergency action' to stabilize the crisis. And it probably did keep me alive in the immediate term. (Though at the time that didn't seem like a benefit.) It allowed me to continue existing in this state rather than deteriorating further. Which is better than nothing and significantly more than most can hope for without a waiting list. But it didn't fix me. It didn't get me better. It kept me from dying but it didn't show me how to live.

For me, the drugs and talking therapies reduced the symptoms and papered over the cracks so that I could appear to function. They offered ways to manage the depression, but nothing to actually make it go away. I had at my disposal every therapy money could buy and nothing, certainly in those first couple of months, was doing much to fix me. It felt like throwing ping-pong balls at castle ramparts and expecting them to collapse. Everything just bounced off.

A typical morning would go like this. The alarm would go off at 6 a.m. and Paul would get up for work. Our alarm is set to 'birdsong' rather than 'buzz' in an attempt to help me wake up un-flustered, which turns out to be futile. If I had taken a sleeping pill, the noise of the slightly mechanical birds would seep through the fog of the drugs and I would come round.

There are a few glorious milliseconds when you first wake. We all have this time, when you are conscious but haven't remembered anything yet. It's like the computer is on but the apps are still opening. You are aware, but your mind is clear. No thoughts. No memories. No emotions. Just you. Being. Sometimes I think maintaining this state would be the highest form of enlightenment – simply to exist, without interference from your thoughts. It is a moment of blissful ignorance.

Then I would realize I was still alive. And I would start to cry. Why couldn't I have died in my sleep? What a bitter disappointment not to have been spared another waking moment. Now I'd have to find some energy to survive the day. Again. The day with all those micro-decisions, all that movement, all that stimulus. It was torture. Bear in mind that my day consisted now of little more than lying in bed, but this alone was overwhelming. All those hours having to keep company with the raging furnace in my head.

As my brain loaded up, the thoughts would start. It's difficult to describe what those actual thoughts were; it's like a firework has gone off in your brain and set every thought you ever had playing on repeat all at once and at full volume. Exhausting.

Some mornings I wouldn't make it beyond the bedroom once Paul had left for work. I would keep the blinds closed and notice the passing of the day only by the change of light seeping through around the edges, the air becoming increasingly stale and stuffy. It never occurred to me to get up and open the blinds, the window, let some life into the room. I didn't want any life. I preferred the cave.

Sometimes I did venture out of the bedroom and go downstairs. Let's not call it anything as grand as 'getting up' though. A shower seemed like an impossible effort. We've already established that the taps were beyond me. And then the decisions that would need to be made were overwhelming. Shall I wash my hair or not today? If I do, shall I wash my face while I let the conditioner sit or will that be too long? Soap or shower gel? Flannel or no flannel? And afterwards, don't forget deodorant, moisturizer. Contact lenses or glasses today? Teeth, Kathy, brush your teeth. These are micro-decisions we don't usually even notice ourselves making, but when your brain stops working properly you have to hold its hand through all the steps, and I simply couldn't be bothered to make the effort. Depression is a double-whammy: it robs you of the mental capacity to do things *and* the emotional will to try.

I did usually manage to dress though. But that's because I had a strategy and had taken to wearing the same thing every day: leggings and a huge cream woolly jumper my mum had knitted. It saved having to decide what else to wear and I ignored the catalogue of stains and drips on the front.

Downstairs, Paul would have filled the kettle and laid the tea things out for me, reluctantly leaving for work and wondering what he might find on his return, hoping that tea at least might offer some consolation. If the prospect of tea was enough to rouse me, I was grateful not to have to think about where the tea mugs were, how many tea bags I'd need, and so on. What is it about tea? It's such a cure. Both comforting and stimulating, it would gently take me by the shoulders and edge me towards the day with a kind word and a pat on the arm. A cup of tea says, '*It's all right, that's it, one foot in front of the other. Off we go…*'

My therapist had given me a worksheet to fill in every day, documenting what I did with each hour, like a school timetable. I loathed her for it. What did she think I was – a child? Plus, this was dangerously close to journalling and therefore reserved only for people who have pens with tassels on the end and own highlighters. Of course, she knew exactly what she was doing and how to push my buttons. I was the school swot. I *always* fill forms in, and I always, *always* do my homework. She had seen the Good Child in me and was going to exploit it in my favour. The knowledge that I had to fill it in made me think about structuring my day. So far it read:

7–8 a.m. – cried in bed
8–9 a.m. – got dressed (it took a long time)
9–10 a.m. – made tea. Drank tea
10–11 a.m. – um…

Fine. I would read, God damn it. That would fill in a blank space in the worksheet. And it's what people do when they are at leisure, isn't it?

I sat down to read *Howards End*. I love this book and I have a beautiful clothbound copy of it that I treasure. I had started re-reading it weeks ago when my brain was only semi-malfunctioning but had not picked it up since. I took to my favourite chair, which is the perfect reading chair – big, squidgy and placed in the corner of the living room by a window. I read a word. Then another word. Then another. I could understand the words individually, but I couldn't seem to join them all together to make a coherent sentence. As the words arrived in my head they were being drowned out by a cacophony of thoughts, so that by the time I'd read the next word, the previous one had been shouted down and I couldn't hear it over the din to link it to the following one. It's like when Spider-Man turns on his Spidey Sense and can hear all of everything that's being said everywhere at that moment. But not in a good way.

And the longer you sit looking at a word, the harder you try to string the words together, the louder the noise gets in your head until it feels like there's a riot inside – angry shouts, fist fights, windows smashed, Molotov cocktails arching through the air and fires breaking out.

The constant chatter in my mind, delusional, self-critical gibberish, was so incessant that I could concentrate on nothing else. I could not sleep. I could not read. I could not watch Wimbledon, make tea, stack the dishwasher, cook, drive, shower, write. The cacophony was overwhelming.

10–11 a.m. – looked at a page of *Howards End*

This is a particularly unkind side of depression. You can't see it. Outwardly, here is a regular person relaxing in a cosy chair, quietly reading a good book. She hardly moves. She might be a picture from a *hygge*-interiors magazine feature (if you ignore the stained jumper). But inside her head the noise is deafening. Those thoughts she can distinguish above the general clamour are screaming that she's worthless, a shameful failure, and everyone she loves would be better off if she died. Her body is fizzing with adrenaline, heart pounding, jaw clenched, stomach lurching in terror. Inside a fire is raging.

My disintegrating mind was completely invisible. I looked just the same. Perhaps a little tired, but that could just have been workload. That was one of the strangest things about those early days of depression when I was still at work. It felt like the world was carrying on and demanding I take part as usual even though, inside, the foundations of my mind were crumbling. My insides were ablaze, my reason infested with paranoia, my sense of self utterly destroyed, my whole world was falling apart around my ears and no one could see it.

If you turned blue or something, at least it would be obvious. And if it was visible the need to fix it would somehow be more urgent. *'Goodness, she's turned blue, this can't go on, we'll have to act.'* Physical symptoms (though I would argue mental health and physical health is a false separation), manifested in some way that people can understand, feel more acceptable than mental ones. And because you can't see it, because it isn't there in everyone's faces all the time, because you can appear 'perfectly normal', it doesn't seem so urgent or acute.

Though that's not to say you can't feel it physically yourself. I was surprised by how much I could feel depression in my body. I have never been thick-skinned, always prone to dwell on criticism or throwaway comments. But now it was as though that internal sensitivity had become physical, like my hide had been removed and I was left raw. Now, every word, every noise, every touch was agony. A door shut too briskly would make me jump as if a gun had gone off. The sun seemed hotter, and I felt like I was being burnt. Lights were blindingly bright. The joyous shrieks of children playing outside sounded like screaming to me. A long-suffering friend sighed sympathetically during a conversation, but I heard an exasperated groan and descended into a spiral of paranoia and disgust – she's right, I'm useless, I'm infuriating, I'm weak. Everything was heightened. I had always assumed depression was a mental experience, but this was very, very physical, albeit invisible.

The worst part about the invisibility of depression is that it makes you question yourself. A broken leg is unquestionable. It's fact. Binary. Look, I can see it on the X-ray. Broken. Real. A mental breakage has no such diagnosis. It's a matter of opinion. It's by degrees. So, is it real at all? My broken mind, prone to questioning and lacking confidence in its own judgement, would question whether I was making it all up anyway. Maybe I'd just worked myself into a pickle, like a toddler working themselves up into a tantrum for no reason. Maybe I was just a narcissistic, over privileged, over-thinking softy who was only signed off work and having therapy because it was an option: I could afford it and I had a partner whose salary could cover the mortgage. If I *had* to work to keep a roof over our heads, or *had* to function because there were children to care for, maybe I wouldn't have time to get into this state and I'd just *buck up*.

But depression is not like breaking a leg. There isn't one point in time when it starts. You aren't fine one day and then depressed the next. It creeps up gradually. Or it did for me anyway. And it isn't as if you feel a bit sad and that sadness builds over time. In fact, the sadness was the least of it for me. What happens is that your sense of what is normal is slowly eroded. Then, after several months, you look at your 'normality', your day, your routine, your habits, feelings, priorities, and realize they have been altered beyond recognition. And it is the depression that has chipped away at it all.

At first you might have the odd night's broken sleep; that carries on for a few weeks and within a month you 'aren't a good sleeper'. That's your new normal. Then, in a dull meeting, your mind flits from one thing to another, unable to complete a train of thought before being distracted by a new, equally inconsequential one. This habit seeps out of the dull meeting room and into every other room. And so it becomes a habit: you have a short attention span. Next you see a driver forced to break quickly to avoid a pedestrian glued to her phone. It's a near miss and you start imagining what would have happened if she'd been hit. Your imagination runs riot, picturing the blood on the road, the distraught driver, the ambulance trying to push through the traffic as you watch on in horror. You can't shake the feeling of impending doom all day. In fact, now you come to think of it, you can't shake the feeling that catastrophe is around every corner at all. Congratulations, you are now a worrier. To add to the worry, there's one negative comment in your annual appraisal, a chink that made it not quite perfect. You dwell on that comment for weeks, picking at it like a scab, allowing it to infect every interaction at work. And, come to think of it, all your colleagues are probably noticing that failing too. In fact, they've probably all decided something has to be done. They are probably all talking about it when you leave the room. You dwell, imagining the conversation with your boss when she fires you, feeding the paranoia. It's normal. This is all normal, for you.

All of these things started as tiny habits or personality quirks that crept in when my defences were low and took hold. But the point is that they grew so many in number and escalated to such an extent that they became overwhelming. If you can never sleep, never settle your mind, never step outside without fear of catastrophe, worry about every inadequacy, well, you don't have much life at all.

Sometimes I thought it might be quite nice to have a broken leg or a back, because at least there would be something tangible. Something that would make it OK for me to be out of action, to be acceptably incapacitated. It would be something fixable too. Doctors could solve it, eventually, and friends and family could help. I could lie on the sofa, legitimately dysfunctional, for a few months. I would take some medicine and do some physio which would fix me and then I would be well again. So tidy, so linear. A broken body would be so much neater than a broken mind, I thought. The high, awkward, sharp-edged flight of stairs that runs up three storeys of our house would be perfect for breaking something and I would stand at the top of them, imagining this neatness, and willing myself to slip accidentally-on-purpose. Then I would spend the rest of the day berating myself for being too weak to do it.

I felt the same about prison too. I listened to something on the radio about a man who was in prison for twenty years. He had studied, read and meditated and become a monkish intellectual. 'How fabulous', I thought. 'No responsibilities, no worries, lots of structure and lots of time to feed your mind with great literature.' But it would need to be in solitary – I couldn't stand the company – and that would mean doing something pretty horrendous. I honestly thought seriously about what was the least-awful thing I could do that would get me in prison, before shelving the idea on account of not wanting to distress my parents. Nonsense of course, but then my brain was all nonsense then.

A coma was another fantasy, but more fleeting. Just to be switched off for a while until my brain had settled down. But comas are unpredictable, and hard to induce yourself, and I was worried I couldn't control the outcome.

Drugs and alcohol would have been an obvious way of escaping. Many depressives turn to them, and I can see why – a momentary release from the turmoil. But I worried about the calories in booze making me fat and, my size having always preoccupied me, that would only add to my self-loathing. Plus, buying drugs in the Cotswolds is a protracted affair involving lay-bys and burner phones, I'm told. Plus, they sat in the same camp as a coma to me – too unpredictable. I had tried cannabis once at university and it made me the most annoying person on earth. I hated my stoned self, and I had enough self-hatred already so that wouldn't help. My worry was all other recreational drugs would have the same effect.

All of these ridiculous schemes, which you can see I considered in obsessive detail, were a way of not thinking about suicide. Because that always seemed like a bad plan. I didn't want to die. Not for ever anyway. I just wanted to be switched off for a while. Really, I wanted all the chatter in my head to be silenced and the grief to be soothed. Sometimes I wanted peace so badly that I thought if death was the only way of getting it, then so be it. But, surely I must be able to find another way of having some respite, even temporarily, from this distress?

I never had a plan, which they say is the critical sign loved ones should look out for: if a person knows how they would take their own life, an actionable plan not 'just' an inclination, then they really are ill. I did think of walking in front of a bus occasionally, but it was only fleeting. If I had really considered it, that's the option I would have chosen. At least that way my family could say it was an accident, I thought. And, in a way, it would have been. Of all the options this one appealed to me most because it took the decision out of my hands. A bridge or a rope will certainly kill you; they required you to really commit to death, to take responsibility for it. But an idle step onto a road is something you do every day. It might kill you, but it might just injure you – fate was responsible, not you. And if it did just injure you, at least you'd be legitimately in-capacitated and have the neat respite you were searching for in the first place.

The crazy thing is that, even as I write this, I think, 'Well if you never had a suicide plan, then you weren't that ill, were you?' As if having a plan means you are a better depressive. I was not a gold card depressive, not a proper one. No free access to the lounge for me.

It all comes back to the neatness and the visibility of depres-sion. If I had been plotting suicide, or attempted it, or seeing things, at least, I thought, it would have been obvious. The de-pression I had was invisible. And if others couldn't see it, maybe I was making it all up. And that made me feel even more alone. And if they couldn't see it, how could they fix it? On the face of it, this invisible manifestation of depression might seem less acute to outsiders, but if you could see what I saw you would be looking out on a lonely and seemingly endless plain of doubt, self-loathing and despair.

And still, it's only 11 a.m. and I have a whole day of this bloody timetable to fill in: the silent oppression of a blank worksheet.

Most days, my mum would arrive by mid-morning and stay for much of the day. At some point it had been decided that I was not to be left alone for very long. We never spoke about it directly, but I think we all knew why.

The antidepressants began to take effect after a few weeks and, contrary to popular belief, they were not kindling happy emotions in my brain so much as pouring cold water over all my emotions, good and bad. In the same way that chemotherapy affects all your cells, not just the cancerous ones, so antidepressants, in my experience, kill all your emotions, not just the bad ones.

Now, rather than fidgety and terrified, I was numb. Utterly blank. No opinions, no preferences – totally indifferent. So, when Mum made me tea one sweltering July day and suggested I get out of the stuffy bedroom and take my tea into the garden, I cared so little either way that I just did as I was told. Which is rare.

I clutch my tea and take up my old familiar spot perched on the edge of one of the raised vegetable beds, just as I had done at daybreak when I was still working. But it is too late in the day for sobbing and the drugs mean I don't feel like it anyway. In fact, I don't feel anything. The sun, high overhead, affords almost no shade; the beds are full of weeds, ramshackle and unkempt from weeks of neglect. I balance my mug on the wooden frame and place my hands on a bare patch of soil. I stare. My mind is blank. I see only my hands and the soil. I feel only my skin in contact with the earth. Both are warm from the July sun.

And there is silence.

Nothing. No internal chatter, no bickering, no feeling at all. Just quiet. It is like a deep inhalation. A drawing in of breath. A preparation. For something. Something is coming.

The exhalation. I feel movement. A tickle. An ant is scaling my thumb. My hand has blocked its route to who knows where and it has elected to scale the mountain range that is my hand. In the valleys between my fingers, I notice tiny, newly germinated weeds fluttering with the breath of my exhale. An armoured woodlouse bustles over the soil like a self-important general off to organize the troops. The soil that I thought was bare is in fact fizzing with activity. With life. Like when you look at the apparently empty night sky for a minute before the stars pop out. My tea goes cold as I watch the soil and the life within it.

12 p.m. onwards – sit in the garden. A change

Our garden is roughly a third of an acre, which, by Cotswold standards, is modest. We built four tiny raised beds when we moved in using railway sleepers and so, back then, I had approximately seven square metres of growing space at home. (We have reconfigured them since; I have laid claim to a bit more of the garden and the raised beds now offer around eighteen square metres of growing space.) At first, this small space was plenty because I was new to veg growing and I had little time to devote to it because of the London job. Plus, I was very precise about it all so everything took ages. I expected perfection from myself, and the veg beds were no different. They had to be flawless. Neat, weed-free and orderly. Like a Beatrix Potter illustration. I spent hours tending each plant individually. I would sow radishes one at a time, measuring a two-centimetre gap between each miniscule seed; the pernickety fastidiousness of a new, insecure gardener combined with my perfectionist tendencies to create a horticultural monster. But as I got more confident it became clear that these raised beds would not be enough to contain my ambition.

One of the advantages of living in the Cotswolds is that many farmers here are hobby farmers, with more land than they know what to do with. My friend who lived five minutes' walk up the lane, The Benevolent Farmer Brown, as he became known, was one such gentleman farmer. He had a smallholding with a walled kitchen garden at the back of the house, bordered by, on one side, the cows' field, a long glasshouse on the other and, at the end, a gate out through the orchard, where the chickens lived, and on to the paddock and Blossom, the sow.

The dry-stone walls of the kitchen garden were weatherworn and crumbling, but within them were four enormous vegetable beds. Farmer Brown only had time to cultivate two beds and the other two, measuring four by eight metres each, were fallow. A couple of years after we moved to the Cotswolds, Farmer Brown suggested I might like to use the spare beds to grow my own vegetables, coming and going as I liked. Sixty-four square metres of soil for my very own use. I tried not to look too wild with excitement as I said 'yes please'. I would grow there for seven years, and it was a dream.

When, years later, Farmer Brown needed his spare beds back to accommodate grand extension plans, I threw myself on the mercy of another landowning neighbour, who showed similar generosity and took me in to grow on his spare patch. Thank heavens for the kindness of country folk with big gardens.

What with everything that was going on that July, I hadn't visited the farm patch for weeks. But as my tea cooled that day and I sat counting the worms in the sunshine with my hands in the soil, rubbing it between my fingers, the loam working its way under my fingernails, I wondered how the walled garden patch was doing. It was the first time I had thought in a whole sentence for weeks.

Something stirred in me that day. I returned to the soil most days for the next week or so. Hours passed as I gazed at the ground. And I kept wondering: is the soil different up at the walled garden? Are the weeds the same? Are the wood-lice as officious as the ones here? Did those lettuce I sowed in a hurry that Sunday afternoon before I flew to New York all those weeks ago ever germinate? I would have to go and see.

Which meant leaving the house.

I had been ferried to and from various doctors' clinics by friends and family, but since being signed off, I hadn't really been anywhere on my own and of my own volition. I wondered if I could remember how.

The five-minute walk up to Farmer Brown's smallholding is picture-postcard gorgeous. You begin on the village green outside our house – a triangle of green with Cotswold-stone houses around. The lanes either side go nowhere much. They are so little used that grass grows up the middle and locals walk their dogs without leads. Take the right-hand lane for tractor-muddied tarmac, precariously fenced sheep fields and deep hedgerows that are thick with elderflower in the summer and sloes in winter. Take the left-hand lane, my lane, up past a row of cottages, more sheep and then sharp right for Farmer Brown's and the veg beds.

The slight incline on the lane before this right turn culminates in a wide vista across fields – a dilapidated grid of low, gappy hedgerows and the occasional spurt of an ancient, ivy-clad tree stump, then on towards the flat expanse south of the A40. On a clear day you can see to the Chilterns and the blinking lights of Brize Norton RAF base flickering in the middle distance – far enough away to be quite pretty.

I walk up the dusty drive to the farm and round the side

of the stone farmhouse to the walled garden. What's lovely about this walled garden is that it is both comforting and expansive at once; contained and exposed at the same time. You are embraced by the stone walls that wrap around you at shoulder height, but you can still see the farm beyond, the sky disappearing over the fields, the woods in the distance. It's so perfectly positioned in the landscape that you can see all of nature, near and far, around you but, at the same time, be sheltered from its buffeting winds and extremes. Cocooned but not claustrophobic. It was worth coming just for this.

As expected, my two beds were messy and neglected. But I didn't care; the space felt familiar and comforting. I didn't do anything once I'd arrived. I just wandered through the weeds and felt the soil. If Miss Haversham had been a grower, her allotment would have looked like this. A mass of tangled weeds covered signs of long-abandoned cultivation. The barely discernible row of bolted leeks, left too long in the ground, were now strangled by bindweed on its quest for height. There were patches of bare soil, so dustbowl dry from the July heat that not even weeds could anchor themselves in it and the rocks and stones of ancient wall repairs had wriggled their way through the friable, structureless soil to settle on the top. It looked more like rubble than earth. The top layer of soil was so crumbly that a breeze could lift it and blow it around the beds, depositing a fine dusting over the bindweed's white trumpet flowers – sooty flecks disfiguring an alabaster devil.

I thought about what this patch used to be like. I thought about the potatoes that had been planted in the end of the far bed the year before and made that wonderful new potato salad – so simple, but so satisfying. I looked at the leeks which had gone to seed and wondered whether I could collect the seed. I sighed at the state of the chickweed that was going for world domination. I searched for the hastily sown lettuces. No sign.

As I pondered, I realized that I had been thinking about the vegetable patch for at least five minutes without interruption. Five whole minutes on one train of thought. No chatter. No critique. No doing. Just thinking. And being. Totally in the present moment, without judgement or expectation. From the outside it might seem like nothing, just an inconsequential daydream about gardening and an idle moment of wandering. But it represented a change. It felt like a tiny life raft, and I clung to it.

Herb Omelette to Soothe

A few herbs did manage to survive in the beds because, unlike the lettuce, they mostly take care of themselves. This self-reliance is why I think of herbs as a gateway drug to veg growing: if you can get people hooked on easy, useful crops like parsley and mint, you can then encourage them to branch out into more unusual herbs like chervil and, once fully addicted, on to vegetables.

Of course, herbs can adorn almost any dish, but put centre stage in a simple omelette is my preferred way to convert people to growing.

For which, whisk 2 or 3 eggs in a bowl with a little salt. Add a handful (3 tablespoons) of finely chopped leafy herbs – parsley, chervil, fennel, chives, hyssop, sorrel... the works – and stir. Heat a knob of butter in a non-stick frying pan until fizzing, then pour in the herby egg mixture and swirl it around the pan. Cook gently, folding in half as it begins to set. Slide the silky pillow onto a plate and grab a fork. As calming to make as it is to eat.

After that visit, I repeated my amble up the lane to the farm patch most days. At first, I didn't do very much when I got there. Sometimes I took a trowel and a bucket to give the illusion of weeding, but really I wanted to absorb myself in the world within the soil. Any ambitions I had had of returning to work after a couple of weeks' rest were abandoned. My advertising life seemed a world away now.

I had found a refuge. An escape to another world, one of bugs and worms, woodlice and slugs. Of mice leaving little piles of gnawed debris from their sweetcorn-stump nibblings; blackbirds jabbing the ground with their beaks to skewer a worm. Of epic battles between a hairy bittercress weed seedling and the pebble it would need to push aside to grow; of sparrows making dust baths next to the battle.

I would watch all of this. Entering their domain and seeing all these struggles and goings-on from on high; their struggles, their wars, their daily chores, like a giant visiting a new planet and watching the tiny inhabitants with detached fascination.

Woodlice, I came to understand, offered particular entertainment. One day, I spent a morning watching as a woodlouse battled to move a fleck of bark that was in its way, her tiny body straining against the weight of it, little legs scurrying beneath to find traction for the next heave.

'Gosh,' I thought, 'that woodlouse is really wound up about not being able to shift that piece of bark. She must be really pissed off. Perhaps it's blocking her route home. Or maybe even blocking her *actual* home. Look how her whole being is consumed by a task that seems to me so insignificant.'

I knew how she felt: totally focused on this work, at the expense of everything else, and completely unaware that the work was, in the big picture, mostly unimportant. In The Grand Scheme of Things, everything that kept me awake at night, that drove me to work so hard, that fuelled the inner critic and created the self-made hell in my head, was just as inconsequential. The world would keep on turning if the woodlouse didn't move the bark, just as it would if I missed a client workshop, or failed to do my best, or let someone down at work. I could fail to be perfect, fail to succeed and it wouldn't matter. Not really. In fact, I could single-handedly bankrupt the entire agency and all our clients with it and the world would still turn, just as it would if this woodlouse failed to move the bark. Physically and mentally, both this woodlouse and I felt like the world was ending. But, from my macro view on high, I could see that for her, it was not. So it probably wouldn't for me either.

I spent a lot of time watching woodlice. It was very comforting to be reminded that what seems a huge deal is, in fact, no deal at all.

The vegetable patch was becoming a daily haven. It offered safety and comfort and peace. An escape. Which was what I had been craving all along. Although I didn't realize it at the time, I had been searching for this kind of quiet comfort for years. It was part of the reason we moved to the countryside; why we took our holidays in rural spots rather than in cities; why I took to growing veg when we moved. We wanted peace.

Even during my most plugged-in, battery-hen work moments I tried to cling to a bit of rural tranquillity to soothe me. When I travelled for work, I packed the *Collected Poems* of John Betjeman. I thought it was because I enjoyed them, but really it was because I was nostalgic for a rural haven I hadn't quite discovered in real life yet – an English countryside that was simple, peaceful and safe.

Maybe the veg patch would be this real-life sanctuary? It was already offering a complete contrast to my London life. Where once I had been surrounded by noise, speed, aggression and concrete all day, there was now quiet, stillness, calm. And life. So much life.

Mum's Fruit Scones

My heroically patient mum, who sat with me day after day that summer, managed to coax me out of the house to the local garden-centre-and-café one day. It sounds mundane enough, right? Like most Cotswold garden-centre cafés it would be full of old, beige ladies drinking weak tea and buying begonias. But for me it was a big deal. So many people, so much noise, so much input. This was exactly what I had been fleeing from when I went to the vegetable patch.

But my mother knows me well, and the prospect of a fruit scone slathered with clotted cream and strawberry jam followed by a trip down the seed-packet aisle stirred something like motivation in me. It wasn't the best scone ever, and I was unbearably anxious, but it was progress.

My mum makes much better scones than this garden centre did. Her recipe is adapted from one in a gnarled, battered and much-loved Delia book:

MAKES 12

40g butter, diced and softened slightly
225g self-raising flour
25g caster sugar

59

ROUGH PATCH

A handful of currants or sultanas
150ml whole milk

Pre-heat the oven to 210°C (yes, really).

In a bowl, rub the butter into the flour using your
fingertips and a light touch until it resembles sand.
Add the sugar and currants and stir them in with
a butter knife.

Pour the milk in gradually, cutting and mixing it
into the dough with the butter knife. Only a butter
knife will work for this. I don't know why, but attempt
it with a mixer at your peril.

Flour your hands, gather the mixture into a soft
dough ball and place it onto a floured surface. Pat it
out to a thickness of 3cm. Take a 4cm pastry cutter
and cut out 12 scones. Do not twist the cutter as you
press it down through the dough or your scones will
go wonky. You will need to reshape the trimmings,
lightly kneading them together and patting back into
a 3cm thickness to cut more scones. Pastry chefs would
baulk at this since overworking the dough can make
it tough. But this isn't Fortnum's, it's fine, really, just
don't handle it too much.

Place the scones on a non-stick baking sheet and cook
for 10–12 minutes or until risen and just brown.

Eat them the same day, which will be no hardship.

August

the scent of fresh
green irises

AUGUST

Immersion

IN THE VEGETABLE PATCH...

Busy but restful. Bounteous but lazy. The August veg patch is all contradiction. On the one hand, the harvests are abundant and require constant picking; the pace of the courgettes is becoming oppressive. If the weather is dry, watering will take up any time not given over to picking or dealing with the gluts. But, on the other hand, there is no actual gardening to be done. All the crops are in the ground by now – brassicas for winter, leeks for next year. It's all done, apart from perhaps another sowing of lettuce, a quick-growing 'catch crop' before the autumn chill sets in.

With only pottering to be done – watering, picking, weeding – just keeping on top of things, the gardener slows down. For an organic grower, like me, there will always be more pottering to do because keeping pests at bay and plants fed falls to the gardener themself rather than chemicals. I, for example, check the netting that protects the brassicas from cabbage whites daily, brushing from the leaves any clusters of, helpfully, neon yellow eggs where a butterfly has managed to get through my defences (they will shred their wings to make it past the mesh). I weed, by hand or with a hoe, weekly and feed (with liquid seaweed or chicken manure) and water the tomatoes what feels like daily because a happy plant is less susceptible to pests and diseases. I have done all this since I started growing, not, initially, for any great environmental ideals – those came later – but simply because the idea of spraying chemicals from a bottle covered in hazard warnings onto something I would later eat seemed, well, just daft. If I didn't need to, why would I?

So the summer chores for organic growers are a little more involved, but there will still be time to lounge in dappled shade, back against an apple tree, lulled by the hazy summer afternoon and, despite the rough bark of the tree trunk, doze; the iridescent hue of tomato vines clinging to your fingers until the scent of fresh green rises, infiltrates the daydream with contemplations of tomatoes, warm from the sun, in a salad for supper.

Make a hole in the soil. Drop a leek in it. Move along eight inches. Make a hole in the soil. Drop a leek in it. Move along eight inches. Repeat. You have a hundred leeks to go.

There will be no space to ruminate, critique past conversations or fret over what disasters might befall tomorrow. The task in hand will take just enough of your concentration to keep your mind in the moment. But it won't be so complex as to make you anxious.

This is leek planting. And it is hard work. The white stem of a leek is unnatural in the sense that it has be created, or forced, by excluding light. To achieve this a grower must sow some leeks in a tray of compost, uproot them when young, then plant each one in a deep hole in the soil leaving only the tip exposed, essentially burying it alive, so that the plant panics and grows towards the light, redirecting all its energy into the stem. This sweet, pale stem is the result of that chlorophyll-free, stress-filled life.

The transplanting stage, when you drop a cherished and fragile baby leek seedling into a hole and wonder how anything so flimsy will survive being buried alive, is all to be done in the height of summer. And unless your soil has been sieved by interns (which I'm convinced is what they do on *Gardeners' World*), then it will be back-breaking work. Like making holes in solid concrete armed with only a toothpick. It will take all day and you will be a sunburnt, sweaty wreck afterwards. Your eyes will sting from sunscreen, your knees will be encrusted with soil, your muscles will be shaking from the effort of getting that darn dibber deep enough into the soil. As you 'puddle in' the leeks, hosing in a splash of water so the soil can gentle puddle in around the leeks, your back aching and your palms worn through from pressing on the end of the dibber, you will look enviously at the leeks in their private pools of water and consider turning the hosepipe on yourself for a moment's respite. The temptation to cut corners and make only shallow holes is strong but skimp now and your punishment will be stubby leeks in winter.

But now the hard work is done. Assuming you have your leeks in the ground by July, as the textbooks recommend, and have not hurriedly bunged them in the ground a month late thanks to your addled mind, then you can just sit back, recover and wait to harvest them in time for Christmas. And when you do harvest them, it will most likely be on a bleak midwinter day when a fecund allotment, rampant with unchecked greenery, seems like a lurid fantasy and when you can't imagine that the sun might ever be strong enough to burn the back of your neck again. This will be when you thank your summer self for having toiled over the leek planting, as you work a fork into the frosted ground to lift a few leeks and carry them to the kitchen for cheesy leek gratin – sweet, silky and cockle-warming.

Leek sowing, like most veg growing, is immersive and absorbing, taxing physically but not mentally. This is part of the reason gardening is used as occupational therapy: predictable, repetitive and immersive, it is everything a tangled mind needs. And there is evidence that it works too. Practitioners call it 'ecotherapy' or 'horticultural therapy', and studies have shown that engagement with one of these programmes reduces anxiety and depression and improves access to employment and social inclusion. There are several charities dedicated to providing horticultural opportunities for people with mental illness. Bridewell Organic Gardens, for example, just down the road from my patch, is a walled garden, vegetable patch and vineyard. Open their wrought-iron gate nestled in the redbrick wall and you walk into a little Eden that is tranquil, beautiful and, rather joyfully, built from nothing by a caring and compassionate community ready to welcome you into their paradise. It is no wonder people find peace and hope here.

There are plenty of cases to demonstrate the success of these schemes and much research to quantify its effect. Though *why* it works is less understood. What is it about horticulture that helps? The connection with nature? The mindfulness it encourages? The exercise? All true. But *why*? The research is not yet definitive. For example, I saw a study which suggests that when you put your hands in the soil, microbes within it stimulate nerves in our bodies which affect serotonin production and mimic the effects of Prozac.

It is my view that the reason science can only quantify the favourable results of connecting with nature but cannot explain *why* it works is because the why is spiritual. I am a card-carrying atheist, and allergic to woo, but when I am immersed in the kitchen garden, I feel myself connected to something greater, something elemental and pure; something that goes to the very essence of life and that I can never hope to grasp, but instead can only wonder at its marvels and find comfort, joy and hope there. Frank Lloyd Wright said, 'I put a capital N on Nature, and call it my Church.' He's right. Nature gives me comfort and a set of values to live by. There is no need for a god here. I have all the comfort I need. This is where my moral compass comes from. And this is real. Not supernatural. This is the Earth. Nature. Life. And remembering we are part of it is a powerful restorative.

I bought plug plant leek seedlings that August. The notion that I might have remembered to sow any from seed that April and had tended them to pencil-thick perfection ready to transplant, in between cowering from commuter trains, hiding in attic conference rooms and taking day trips to New York, was laughable. But I managed to find some cut-price plug plants online and dibbered them into the solid August-baked soil. I was a month late, but that they made it to the soil at all was a minor miracle. I couldn't make an omelette, but this ritual of going to the patch every few days was a routine I didn't have to think about. I just plodded up the lane and did the labour. It was a sort of meditation. I had one job to do and it was uncomplicated and satisfying. There were no decisions to make, just the quiet monotony of dibbering. But it wasn't mindless. It was hopeful. Hope that the seedlings would grow and, though I could never have articulated this at the time, a vague sense that I should hold on for when they did, because a harvest would show me that things might be positive in the future. There was good to come. In the meantime, I was utterly in the moment. And the moment was at the centre of all things – calm, hopeful, healing.

Which was just what I needed after too many years on the

hamster wheel. Back then, imagining myself to be a black-belt multitasker, I thought nothing of listening to a conference call in my earphones while on the train to London with my laptop open so I could make changes to a PowerPoint between gulps of scalding hot First Great Western tea and regular checks of my BlackBerry. (Thank God Twitter and Instagram were only fledgling distractions then.) I never spilt the tea and the people on the conference call never knew I was doing three other things. I thought I was winning. But I was living in a constant state of not quite giving anything my full attention, of trying to hold too many things in my mind at one time, like a tired old computer with too many programmes running simultaneously. It is no wonder my mind ran out of RAM.

But life was much simpler in the vegetable patch. I had just one task and nothing to distract me. I found these jobs reassuring too. I would imagine all the other vegetable gardeners over hundreds of years who had dibbered and puddled-in their leeks just as I was now. It was like stepping out of time, doing work that had been done for years in a landscape that had looked this way for generations. It was a world away from my life in London where things were constantly in flux, changing and splintering with every knee-jerk reaction. Here, with my leeks, I felt connected to the landscape, to the soil, and to the generations of growers who had gone before me.

And connection is what vegetable growing is all about. You cannot help but feel connected to the space. You spend so much time at close quarters with the earth that you develop an intricate knowledge of its habit: which bits are clay; which are free-draining; which are especially infested with bindweed; which are unusually rich in manure because you dumped the final bucket there last season and couldn't be bothered to rake it out. You come to know every inch of the soil, its *terroir*. Which is much healthier than knowing every inch of the 06:32 to Paddington (there is, since you ask, a questionable stain on seat 43 in coach C, and the regular commuter who snores loudly usually takes the aisle seat halfway down coach F...).

This sense of connection and immersion was particularly powerful because of what I was immersed *in*. Some people find solace from immersing themselves in music, or a good book. These are an escape that can transport them elsewhere, which can be helpful in its own way. But with veg growing, what I was immersed in was *life*. The here and now, not elsewhere. Life at its most fundamental, most basic. With the journey of every plant from seed to plate, I was witnessing, at the closest quarters, the very essence of creation.

There are two moments in the garden that I think are the most healing, and they occur at the beginning and the end of the growing journey. The first is the moment you notice a seed has germinated. But you never see it if you go looking for it. I would go out almost daily at first to check on newly sown seeds, my desire to be a success still very much alive and kicking (in steel-toe-capped boots). I would stare at the soil. Nothing. If the seeds never germinated I would notch it up on my extensive list of failures in life. This is what depression does. It makes you see only the negatives, attributing things that haven't gone your way in life as a sign of your own personal failing, part of a pattern, a narrative that you generally aren't as good as you should be.

'Kathy couldn't get her lettuce to germinate. Which just goes to show how useless she is. It's typical really, she often falls below acceptable standards. She's a pretty rubbish person, actually. Best to write her off.'

Clearly, lettuce-growing abilities do not determine a person's worth. It's nonsensical. But in a low, everything was my fault. Things escalate when you're depressed.

Imagine, then, that amid this ludicrous level of self-loathing you wander into the garden one day to find a single fleck of green in the vast expanse of the raised bed. It must be a weed, your critical mind tells you. You look closer. The fleck, a thin pale green stem is still hunched over, half unfurled, having only just emerged in the last few hours, protecting a pair of thin, delicate leaves curled underneath. Their flesh is newly minted, still clean and glossy. This tiny, gossamer shot of life has, with heroic strength that belies its size, grown through the rubble of soil it was buried in and popped up right by the string marking the line where I sowed some lettuce. I look along the string to find that it is not alone. There are two, no, three more pioneers with the same shaped leaves, just breaking the ground. There's no denying it. The lettuce has germinated.

I yelp in surprise. Lettuce! Creation! Yesterday there was failure and hopelessness; just a lifeless row of stubborn seeds;

more bagged-and-catalogued evidence of my worthlessness. Today there is life. And inside I feel... I don't know. I've not felt it for a long, long time. There's a bubbling in my chest that isn't panic or terror or grief. For a moment there is something else. What is it? Is it ... hope?

There, in that moment, in that little green speck of life no bigger than an eyelash, I find all the light I thought was gone from the world. In that single moment I am joyous. I feel pure delight. Childish wonder. I am genuinely amazed at the incredible resilience and ingenuity of nature. What a powerhouse that tiny seed is to have been buried under a mountain of earth and, without aid, found the energy and nerve to grow. It's like dropping a sleeping toddler in a field and expecting it to feed, clothe and educate itself.

It's not that I saw the seedling as a metaphor (this isn't a motivational office poster from the nineties). It was that I felt like I was looking nature – at her most pure and private – full in the face and seeing there hope and comfort. It seemed like I had been allowed to see the inner workings of the world, what went on when you looked past the manufactured nonsense and clutter of constructed life and saw the true character of the earth. And what I found was that the true character of the world was joyful, nurturing and hopeful. Here was an antidote to the darkness in my head. Here was where I needed to be.

Hooked on the elation at that moment of germination, I became immersed in the plot, sowing whatever I could wherever I could find soil that might take it. Despite it being woefully late in the season to start growing vegetables, I ploughed on. Frenzied. Anything to see that first fleck of green shoot out of the soil and show me that there was still life and hope and joy in the world. I sowed seeds from out-of-date packets, random unmarked envelopes of saved seeds found in an abandoned trug in the shed; I liberated herbs from super-market cut-and-come-again pots, divided them and planted them up; I bought a forest of plug plants online, forgetting what I'd bought and from where until they arrived in the post weeks later and I had to find a corner of soil to host them. I couldn't get enough. And it was healing.

The second moment that I find equally restorative is when you come to eat the thing you sowed. Here before you at the table is a lettuce. It might be just a few leaves, but they are big and blowsy and they fill the plate. You've been careful not to do too much to them. Just a pinch of salt, a dash of oil, a squeeze of lemon – nothing that will distract from the mirac-ulous lettuce. You eat the first mouthful. It is crisp and but-tery sweet.

What you are eating now did not exist four weeks ago. Think about that for a second. It has been brought into being because you put a seed in the ground. A tiny seed the size of a pinhead that has converted basically nothing but sunlight and soil into a fat, juicy lettuce the size of a football. You are eating sunshine. And earth. It's awe-inspiring when you really think about it.

Dressing a Lettuce

The first lettuce harvest wants nothing more than extra virgin olive oil, salt and a squeeze of lemon. It is too precious, too extraordinary to require further adornment. After that, as the harvests become more regular, you may like to try pouring the following into a jar – 3 tablespoons of your best olive oil with 1 tablespoon of apple cider vinegar (or, if you can find it, something sweet like apple balsamic vinegar), adding a teaspoon of Dijon mustard, a half teaspoon of honey, a pinch of salt, a tablespoon of hot water – and giving it a good shake. Add grated garlic and chopped herbs as you see fit.

Pour some of the dressing into a serving bowl, add the lettuce and toss with gentle hands. Go easy on the quantities though. Nothing worse than a drowned salad.

Come the second mouthful of lettuce, I'm thanking nature for her super-powers. I am profoundly grateful for them. Now, there's a lot of woo talked about gratitude. 'Grateful' and 'blessed' are toe-curling, meaningless clichés to me. When we are encouraged to list the things we are grateful for (as we inevitably are by some Instagram guru using maybe, God forbid, a journal) our tendency is to rattle off obvious reasons to be cheerful, which, I think, helps no one unless you are already cheerful. I know I live in a beautiful home, have all my limbs and a loving family. But that doesn't make me feel better. Because I have depression. That's the point of depression. So, counting your blessings in this way is futile because when we see things like love, health, nature in the abstract, it's too broad and wishy-washy to have any emotional effect.

Not so with vegetables. What happens when you eat something you have grown is that you make that abstract manifest itself. Here on your plate is a very pure example of the earth having given you something from nothing. It is a process you have watched from beginning to end. You are, in this mouthful, taking part in the most basic and fundamental exchange between humans and the Earth, one which has happened for millennia. It occurs countless times a day, all over the world, but, at its core, it is all you really need for life. And you have just been able to witness it in microcosm from start to finish.

By the third mouthful I'm starting to take some of the credit for this lettuce success. 'I did that.' Except, obviously, I didn't, as we've just established. But I did have some agency in it. I tilled the soil. I planted the seed. I brought the water. I have been nursemaid at the very least.

And a sense of agency is something I've been lacking lately. On the darkest days I have been unable to do virtually anything for myself. I have had to be helped out of bed, put in the shower, fed toast, so heavy has been the lead weight in my chest. And here I am now, with a lettuce of my very own.

A neat, pert, clean lettuce, whole in my hands and ready for the plate. It brings a deep and profound sense of satisfaction. I have made something. And it is beautiful. And useful. I used to get a warm glow from a well art-directed PowerPoint presentation (which I also thought was beautiful and useful). My sense of fist-pumping gratification came from selecting just the right train carriage to place me right at the exit of Paddington tube station. Now I'm getting the same fix from a plate of lettuce. Things change.

Immersing myself in the growing of a seed to eat gave me a connection to the world that I had lost. When I saw nature up close, I knew that this was real life, not the construct I was part of in London.

The priorities, the values, the power structures I had got so caught up in were all artificial, just flimsy inventions that really only exist in our collective mind. I had thought this life was real life. But in fact, the veg patch, nature, was the only *real* life. This was the centre of things, not the bubble I had escaped from. And at the centre of things, everything was simple and clear. In the kitchen garden I was reminded that the worlds we live in day-to-day ultimately do not matter very much, and that what is real is the nature that surrounds us. Real life is in nature. This is where I belonged. This is home. All that, from planting leeks. Who knew what else the patch might teach me.

Mushroom, Chestnut and Leek Pastilla

When leeks do finally arrive, this is what I cook first.
It's a vegetarian version of the traditional North African
dish made with shredded meat, usually pigeon, eggs,
spices, icing sugar and flaked almonds. Yes, an unfamiliar
combination to some, but it really is the business, and
the flavours work beautifully with the sweetness of the
leeks and chestnuts against the deep savoury umami
of mushrooms to make a dish that is both comforting
and novel.

SERVES 4 GENEROUSLY

50g butter, plus extra, melted, for brushing the filo
300g mushrooms, roughly chopped
5 large or 6 regular leeks, finely sliced
3 eggs, beaten
Small bunch of parsley, finely chopped
150g pre-cooked chestnuts, roughly chopped
4 tbsp flaked almonds
2 tsp ground cinnamon
1 tbsp icing sugar
6 sheets of filo pastry

Melt 40g of the butter in a frying pan and fry the
mushrooms for 3–4 minutes over a medium-high
heat. Turn the heat down, add the leeks and a pinch
of salt and cook gently for 10–15 minutes until soft
and sweet but not browned. Set aside to cool.

Meanwhile, melt the rest of the butter in another pan
and scramble the eggs very gently, removing from the
heat while they are still runny. Add plenty of salt and
pepper and the chopped parsley, then set aside to cool.

In a bowl, mix together the vegetables and the
scrambled egg. Add the chestnuts, almonds, cinnamon
and icing sugar and combine. Check the seasoning –
it will almost certainly need more salt. Leave to cool.

Pre-heat the oven to 180°C.

You can make the pastilla freeform or in a tin. For the
latter, brush a 20cm loose-bottomed cake tin with a little
melted butter, then line it with a sheet of filo. Brush the
filo with more butter and place another sheet on top at
a 45-degree angle to the one below. Leave some filo
hanging over the edge of the tin. Repeat with 3 more
sheets of filo. Fill the tin with the pastilla mixture. Place a
sheet of filo on top of the filling, then lift the overhanging
filo sheets up over everything, scrunching them up
as you go to create a ruffled top that seals the pie. For
the freeform option, lay 5–6 sheets of buttered filo on
top of each other at 45-degree angles, pile the filling in
the middle, then bring the edges of the pastry up over
the filling and scrunch together messily at the top, like

Dick Whittington would arrange his knapsack, or like wrapping Christmas pudding. Brush with more butter and bake for 30–40 minutes until the filo is golden brown.

Once golden, remove from the oven. Release from the tin, if using, and serve with a crisp green salad.

September

this tiny miracle,
this alchemy

Solitude

IN THE VEGETABLE PATCH...

My favourite month. The schools have gone back. The shops are stocked, unbelievably, for Halloween. People mention the clocks changing. They have called time on The Summer. Gone back to work after a lazy August to glaze over at their desks and dream of toffee apples. But the weather is still good, often more reliably warm than June, and the harvests – basil, tomatoes, aubergines – are redolent of Mediterranean holidays. Al fresco lunches require only a thin jumper and are colourful, blowsy salads of barbecued veg scattered with mozzarella and torn herbs. It's as if everyone else has left Summer behind and you have it all to yourself now. A secret encore for those who know nature well enough to stay until after the credits.

By the end of September, the patch reaches its peak of product-ivity in the brief, glorious moment when the summer and au-

tumn harvests overlap. Perhaps the peas are over, but the beans, courgettes, sweetcorn and tomatoes particularly are delighted by the long summer they've had to ripen and they remain abundant. And now the autumnal hues of deep blue-green and burnt ochre – beetroots, carrots, kale, turnips, cavolo nero – join them to create a rainbow of plenty in the harvest trug.

Company is a glib, and flawed, remedy to melancholy. Surround yourself with friends, they say. It will cheer you up, they say. Rubbish. Of course it helps to know that people care about you. But suggesting you might be able to dupe your mind into cheeriness by 'getting out and socializing' is tantamount to saying 'buck up'. Believe me, if it was that simple don't you think I'd have done it? For me, at least, interacting with people just opened a Pandora's box of shame. You see your situation through their eyes and the reviews are not good. What must they think of me? Here is a woman with everything she could ever want in life – love, success, money – and still she isn't happy. Just an overprivileged, ungrateful brat who doesn't know she's born. I must be repugnant to them. Being around people didn't console me; it just made me feel ashamed, pained for the ugliness and disappointment I was inflicting on them. It was like trying to hold a conversation with a nun whilst naked. Excruciating.

It was practically testing, too. I couldn't drive and my memory was unreliable, so getting out to visit friends was logistically tricky and dependent on me remembering our appointment. Fortunately, when we moved in, I'd become close friends with a woman who lived opposite us on the village green, a person of almost bottomless empathy and undiminishable spark with so few judgemental bones in her body that even my paranoia couldn't project my shame onto her, who would coax me over to her garden for a cup of tea. She would message in the morning, a casual but authoritative 'Come for tea this afternoon?', and I would have to set an alarm on my phone to remind me. When the alarm went off, I would jump in surprise and check our text conversation to make sure I hadn't imagined it.

And so, the solitariness of the veg patch called to me all the more. I don't think it would have been the same if I'd had a patch in a field of allotments – too many judging eyes. What was unique about my patch was that I could be solitary, but I was not alone. There was company. But it had no capacity to judge. The horse in the neighbouring field would trot over as I arrived, always inquisitive; the farm cat eyed me from its vantage point under the shade of the wheelbarrow (actually, that cat, like all cats, *was* pretty snooty); the rabbits kept their distance but coveted the lettuce seedlings from the safety of the orchard beyond the gate. In the vegetable patch I could be in companionable silence with other creatures without the agony, as it was to me then, of having to worry about what they thought of me or what they might expect from me.

It is September. I am in the farm vegetable patch weeding. I'm on my hands and knees picking out hundreds of minute bittercress weeds before they turn into ground cover. It is fiddly, close work and my nose is barely twenty centimetres from the soil. I am totally lost in the task. I took sleeping pills the previous night so I am a bit spaced out as well; their hangover lasts well into the following day.

As I weed, I begin to imagine I am being watched. 'Don't be ridiculous,' I say to myself. 'You are being paranoid. Please let's not descend down *that* route of madness.'

But I can feel something nearby. There is silence so I must be alone. But I have an overwhelming feeling of eyes staring at me, a presence, menacing, unmoving but intent.

A snort down the back of my neck. I scream, leap to my feet and spin around to be faced with the long lashes, watery eyes and stubbly pink nose – of a cow.

It blinks. Unflustered by my outburst. And carries on watching me.

Once my heart stops galloping, I settle back to my weeding and the cow settles back to leaning over the stone wall to watch me weed. She didn't move for nearly an hour. And I enjoyed her company.

Sometimes, when people ask me now what they can do to support a friend with depression, I suggest they might try to be like that cow – placid and reliable, offering no unsolicited advice, or critique, without judgement and without trying to fix things. Just another creature to share the world with in quiet activity.

Depression, in my experience, makes you self-centred. 'Self-absorbed' might be more charitable, but either way the magnitude of your own thoughts and distress leaves room for little else. For a start, you are so bound up in your own pain, so consumed by it, that you can feel no empathy for anyone else's suffering. You have no room to feel sad for others. All the space is taken up by sadness for your own tangled mind. Everyone else's anguish is subordinate to yours. And not only are you unable to offer empathy, but you are also unable to receive it either. No one else can possibly appreciate your situation. No one has suffered like you. After all, who could feel this awful and live to tell the tale? You must be the only one dealing with these symptoms. Surely no one would have survived it.

In this state, you are terrible at dealing with people. Cows are much easier. If anyone asks you a favour or expects something of you, they are heartlessly demanding. If someone gets in your way or unwittingly does something to make life difficult (I'm talking things as inconsequential as a badly driven supermarket trolley), they are grossly inconsiderate. How dare they! Don't they know you have the worst, most difficult life imaginable? How could they be so selfish as to intrude on your suffering with their petty needs? And if, heaven forefend, your husband delicately suggests you might need to slow down and take a moment, he is gaslighting you into thinking yourself incapable. Outrageous. Everyone else is outrageously unreasonable.

But it extends beyond the usual hassles of life – trains being late, deliveries not turning up, husbands trying to take care of you. Things unconnected with you, not in the least bit even directed at you, become part of the conspiracy. Everyone is out to make your life difficult. To somehow belittle your feelings.

Walking through Paddington station on my way from work once, a tourist ahead of me couldn't find her ticket to get out of the underground. I barged past her, huffing loudly and took her place at the ticket barriers.

'Jeez, like, rude,' she whispered to her friend in a Californian drawl.

And my reaction was not 'Oh-God-she's-right' remorse. I did not look at myself and question my priorities. I did not apologize profusely for my bad mood and show her how to use the barriers.

No. Instead, I thought, 'Bloody Americans, so inconsiderate. Are they *trying* to make me miss the train?'

Of course, all of this was about projection. Me projecting my own lack of self-worth onto others and having them, unwittingly, play it back to me. I interpreted all these small 'outrages' as examples of an uncaring world because I felt I wasn't worth caring about. That I was livid to be treated like this might suggest I thought I deserved better, but really it was just a warped mechanism for underlining that they were perfectly right to treat me with such little care. I didn't deserve it. And all of these 'examples' just proved that.

However you choose to psychoanalyze it, depression made me selfish, myopic, grumpy and insular. It is not a very edifying condition.

(I go pretty easy on people who barge me on the underground these days. They probably aren't evil. They are probably having a really rough day.)

Visiting the veg patch that September, my grumpy, isolated state was less acute than it had been on the underground, but I was still wrapped in a thick, dulling coat of loneliness and had a lingering view that the 'outside world' wasn't welcoming. I thought that in the patch I was protected from the uncaring world. Here, I believed, was the only place where I was secure from its ridiculous demands. I could be on my own. (Apart from the cow.) Which was the only safe option.

But the veg patch had other ideas. The result of my sowing frenzy, coupled with some luck with the weather, was a very incoherent, but prolific, selection of vegetables. The garden was endlessly generous, joyfully producing great gluts of courgettes, tomatoes, beans, lettuce, sweetcorn, cucumbers, basil, aubergines, kale, onions, edible flowers, beetroot, carrots, spinach, parsley and fennel. Even the leeks had survived their transplanting.

I had planted haphazardly and without much thought, just

desperate to get my fix of that elation I felt when something germinated. It was as if I had forgotten that life could be created and lived with joy and purpose – and now I was rediscovering that with each and every harvest. Look! You grew into a sweetcorn. Life. It's true. Amazing. Show me again…

Even if your patch is a little neglected or you are new to gardening, you cannot fail to have gluts. For example, maybe you only grow two cucumber plants. At some point in September, they will both be at their peak and there will be two or three cucumbers on each plant, all ready to eat. And if you don't pick them now, they will turn fat, leathery and tasteless in a few days. So, rather than face the challenge of eating six cucumbers in a week, you give some away. Even if quantity doesn't drive you to sharing your harvest, then the timing of it will.

And what gluts I had that year. More than I could ever hope to eat myself. I had no choice but to share them. To face the 'outside' world, which I continued to think of as hostile. But wasting my harvests wasn't an option. I had no choice but to be generous.

Cucumber and Gin Sorbet

Gluts mean you tend towards recipes that use a lot of something. With a major cucumber glut, a cucumber sandwich won't make much of a dent. But a sorbet will.

Put 300g golden sugar and 40g liquid glucose (which you'll find in the baking aisle) in a saucepan with 300ml of water. Set the pan over a low heat until the sugar has dissolved. Do not let it boil. Remove from the heat and add the juice of half a lemon.

Juice 4 large cucumbers (about 600g). If you have one, you can do this in a juicer. If not, like me, roughly chop them then whizz in a blender to make a watery purée. Line a sieve with muslin and set it over a bowl, then tip the purée into the sieve and leave for a few minutes so the juice drips through to the bowl below. You can gather the muslin up into a bag and give it a gentle squeeze to encourage any last drops. Either process should result in approximately 500ml of juice.

Mix the sugary liquid and the cucumber juice together. Add 140ml gin and put it in the fridge to chill. Once chilled, churn in an ice cream maker then transfer to the freezer to set completely.

(If you don't have a churner, a coarser, more granita-style ice can be made by freezing the liquid for 2 hours in a wide, shallow tray, then roughing it up with a fork. Return to the freezer for 45 minutes then repeat the forking-up process 3–4 times more. The result will be a boozy, bright green Slush Puppie that will keep in the freezer and retain its crumbly texture.)

At work, we had this thing we called a Shaky Trolley on people's birthdays. A hostess trolley piled high with Krispy Kreme doughnuts, Haribo and crisps and clinking with champagne and beer, it was brought round to the recipient's desk (God forbid they should have actually *left* the desk to enjoy their treat) and we'd all binge on the bounty of sugar and booze. It was very convivial, though I suspect we were all self-medicating. You could hear the chink of bottles coming down the corridor as the rickety trolley rattled over the carpet tiles – hence the name – and we would end our conference calls quickly or finish dashing off an email to get to the fizz while it was still cold. The Shaky Trolley also made appearances when a pitch or an award had been won, or on some vague excuse created by management, like it was a Friday afternoon. There were similar trolleys at most of the companies I worked for. At one agency, the trolley got cut and we all knew budgets must be so tight that the end was nigh. A lot of people called their headhunters that afternoon.

Now, in the garden, I had a green, steel-mesh trolley with drop-down sides and a soft-grip handle that steered the front axle. It was my own, much healthier, version of the Shaky Trolley and it had just the same alluring quality that drew people to it. I would wheel it up the lane from my house to Farmer Brown's loaded with the weeding bucket and a spade, then wheel it back full of the September harvest.

Back and forth I went every few days with that green trolley. And it was just as much a social magnet as the doughnuts and fizz-filled trolley, whether I liked it or not. As I trundled across the village green with my harvest (so idyllic, why on earth wasn't I happy?) dog-walkers would pass and comment on how pretty it looked. And neighbours would pop out to raid the trolley of its gluts.

I was seeing people. And talking to them. I was not wild about either. But I had a trolley full of food every week and a village green of neighbours eager to help with the glut. The vegetable patch was gleefully forcing me out of myself by foisting its bounty on me and making me share it; bowling me over with the sheer weight of its abundant, generous harvests knocking me out of my isolation.

I'm not sure you could say I was chatty, but that I could talk to the people who stopped at all was progress. We would discuss what to do with the vegetables and chat about the weather. Nothing profound – the antidepressants numbed any capacity for intensity – just normal, everyday living. And here I was, *living* too. Alongside taking myself to weekly therapy sessions, rather than being driven, and the elongation of time I might safely be left alone, I now added small talk to the list of 'normal' things I could do. I reported all this to my ever-supportive bosses as evidence that I would soon be back but was met with a firm no. They could see this was progress, but I was far from well.

Most importantly, my harvests meant I had something I wanted to give the people I spoke to on my new 'commute'. Mainly marrows, but still. I wanted to share. I had discovered this tiny miracle, this alchemy by which a handful of seeds could transform into a trolley full of food, and it was transforming me too. Did everyone else already know? Had I just missed the memo? Had they already been unplugged from the Matrix and were also now able to see the real, true workings of the world? Did they see this marrow? It was a seed three months ago and look what joy it was bringing us now. What potential. What hope. They must take it.

It was the first time I had noticed any generosity of spirit in myself for many months. The harpy who terrified tourists on the underground was being exorcised by, of all things, the marrow glut.

I still have that green trolley and it still carries my harvests. It is a bit faded, one tyre is split (and, as it turns out, impossible to replace), the soft-grip handle has disintegrated, and the drop-down sides drop down randomly and of their own accord, but I can't bear to part with it. I feel real affection for that trolley, like you do for your first car.

Marrow Jam

The only thing difficult to shift from a bounteous trolley of vegetables, is marrows. However fervently you foist them on passers-by, everyone seems to have a good excuse not to take them. Reasonably enough, since they don't make great eating – tough, watery, a touch bitter at worst, bland at best. They need help. And jamming them is their redemption.

Grate 1kg of marrow, skin and flesh, but not the big seeds at the core, into a jam pan or large stainless steel saucepan. Set the pan over a medium heat and cook, stirring regularly so it doesn't catch, for 5–6 minutes or until the water starts to run from the marrow. I have known wise jam makers add 3–4 teaspoons of grated ginger and the juice of a lemon at this point too.

Add 1kg jam sugar (i.e. with pectin), stir until it's completely dissolved and then turn the heat up high and bring to a rolling boil, where it must sit for several minutes or until it reaches 104°C (for those with a jam thermometer). Ladle the hot jam into hot, just washed jars (you'll need 3–4 jars as these quantities make roughly 1.2kg of jam) and seal. Assuming you have scrubbed the jars sufficiently, and both jam and jar are hot when filled and sealed, the jam will keep in

a cupboard for months, but is ready to eat as soon as
it is cooled. Store in the fridge once opened.

Now, don't imagine that one day I was a grumpy recluse and
the next a giddy, marrow-pushing socialite. What I wanted
to share was the quiet joy I had found. Quiet being the key
word.

In the veg plot, I might go for several hours without speaking. Now that the internal chatter and constant criticism in
my head was quieter, I luxuriated in the silence within. Like
when someone turns off Radio 4 and the air in the room re-
laxes. My mind didn't fill the silent void with other ramblings
– no To Dos or 'I mustn't forgets' or 'what on earth am I going
to do with my lifes'. I was so immersed in the soil, so busy
with growing that my internal fretting was drowned out by
nature.

Like meditation, I was perfectly in the moment. I observed
the quietness in my head, noticing how my mind took on a
fluid quality when I didn't have to form sentences. Thoughts
might come and go but I felt no need to order them or engage.
My task was here in the soil. Now. No distraction. Peace.

I wouldn't come to fully understand the restorative power
of spending time in silence within nature until many years
later, when I went on a silent nature retreat.

I had started meditating when I was signed off work after a
friend recommended the Headspace guided meditation app.

At first each session was little more than sitting for ten minutes and listening in horror to the cacophony of my dismantled mind. I lapsed frequently, coming back to it time and again over several years, messily and inconsistently, until I could consider myself well versed in all the techniques, though, if I'm honest, never feeling profoundly changed by the experience.

Quite why I decided to try a three-day silent retreat, I don't know. I argued that I wanted to see what happened when I removed all stimuli. Where would my mind go when I unplugged – no phone, no pen, no books? This was several years after my first bout of depression, but life was starting to get in the way again and I could feel the pressures building. I am much better at spotting the signs these days, so I knew I needed to act quickly before things got out of hand. Constant vigilance – the motto of any recovering depressive. Plus, the stately home where the retreat was held looked fabulous. And, best of all, it had a kitchen garden.

The garden is the first place I go to when I arrive.

The metal gate is straight out of central casting: picture-perfect secret-garden material – battered, rusty, not quite on its hinges thanks to the climbing wild clematis that is slowly conquering it. Fully in character, the gate creaks as I open it. Not a little squeak. This is no am-dram gate; it's a pro and probably practises this performance daily. It's a proper cartoon, full-volume, haunted-house creak, oozing mystery and drama such as would make Karloff proud. Ordinarily I'd be delighted that such textbook gothic charm should announce my arrival into a walled kitchen garden but today I cringe because this creak is the loudest thing I've heard all day, and that probably also goes for the twenty other people hidden around the estate whose peace I have just shattered.

I wince my way into the garden, the whole melodrama having been repeated as per the sign instructing me to 'please close the gate', and instantly all is forgotten. I am transported. Rows of green beans, tangled and riotous, stretch before me. Battalions of kale trunks are tucked neatly under butterfly netting. A sweet shop of colours makes up the cut-flower beds, yellow fennel flowers lolling against a tepee of sweet peas as *Alchemilla mollis* quietly takes over underneath.

Everywhere I look is dripping with pumpkins, courgettes, beetroots, peas, leaves of every colour. And weaving through it all are nasturtiums, calendula, borage, left to run wild like naughty, gleeful children. In the distance, against the long brick wall that borders the garden, I can see a vast lean-to glasshouse struggling to contain a jungle of tomatoes, aubergines and cucumbers. It is heaven. Sheer exuberant, effortless abundance. I think I might cry at the beauty and joy of it all.

Taking my shoes off, the better to enjoy the grass paths that mark out the beds, I pad my way to a bench at the end of the dahlia beds. The bench gives me a perfect view of this glorious garden and a hint at the menu for my next few days, since virtually all the food comes from this kitchen garden. From my bench I can see just how much is packed into this space.

I am bowled over by the productivity. It is bigger, more tight-ly packed and better run than my patch. But I feel at home among these vegetables. So, I decide to stay for a bit. In fact, I do not move for two hours.

Two. Hours. Can you imagine it? Just sitting. For two hours. Have you ever actually done that? No, me neither. Not un-til then. Nothing to do, nothing to achieve, nothing to think about, no phone/book/pen/humans to distract. Just nature.

At first, I think the only sound I hear is the vegetables rust-ling in the warm breeze and I am thankful for such quiet com-pany. But gradually I realize that this place is anything but silent. Blackbirds jab at the soil, flicking it aside in their pursuit of worms; something unidentified ferrets about in the under-growth; a squirrel, claws clattering on the bricks, scuttles along the wall of the garden with cheeks stuffed full of contraband. Then, and I swear this is true despite how implausibly idyllic it sounds, a rush of wings to my right and a robin comes to rest on the back of the bench and enjoy the view with me for a moment before flitting off towards the herb garden. This is not solitude. And nor is it silent. I can hear wildlife, plants, weather – all of nature. It is a cacophony.

Over the next three days here I come to realize something I had already observed in my own veg patch but had never man-aged to articulate. That when we are silent and remove all the other noises of human life – phones, chat, people, the million tiny decisions we have to make every day that provide a con-stant chatter in the mind – when you remove these things, the sound of nature comes to the fore and her noises seem heightened, more intricate, full Dolby Surround Sound. And this is very restorative. It makes me feel connected with nature but, most of all, it is loud enough to drown out the din of daily life. With this natural soundtrack in my head, there is no room for other thoughts, worries, conjecture, lists or ponderances about what I should/ought/must have achieved by now.

But the non-silence of nature doesn't simply distract from the chatter in one's head. It reframes it too. It's as if the rustle of the trees, the woosh of the swallows, the gentle munching of the cows in the field, the ferreting of the mouse among the leaves, are joining together, and nature is shouting, '*Here I am*. This *is the world*. This *is real life*. This *is what you find when you put down all that gadgetry and made-up constructs you humans hem yourselves in with and look up instead*. Here *is what's real*.'

Some time later, on another silent retreat (I went back several times, couldn't stay away), the retreat leader spoke (he was allowed to) about the Buddhist notion of *sati*, which is commonly translated from Pali into English as *mindfulness* but more literally means *remembering to observe*. I thought this was a perfect way to describe what I felt when surrounded by nature in this way: like I was remembering to see what was real. I defy anyone not to find clarity in the beauty of nature, any nature, if they just sit with it in silence for an hour; not to feel more aware of, more part of the Earth.

When you surround yourself with nature in this very particular, silent way, giving yourself time to just sit and absorb it, to get *bored*, you realize that you are part of nature too. You are just a creature, same as that fat little squirrel on the wall. So is that other retreat guest over there. Or that sheep on the hillside. Or that heron by the river. We're all just beings. And all we're doing is just, well, *being*. That's it. That's all there is to do. Just *be*. Everything else is a fiction.

Courgettes Stuffed with Lamb, Harissa and Apricots

For all the life lessons the vegetable patch has taught me, there is one that I am still to learn, despite it being demonstrated to me several times: you only need a couple of courgette plants. One plant will happily feed a family of four all summer and yet I plant eight at least. It is a compulsion. So, every year, courgettes are my biggest glut. Fortunately, I have become expert in what to do with them and stuffing is my favourite method.

And you will not go far wrong with a lamb mince stuffing. The symmetry of nature is such that, during the September I was ill, The Benevolent Farmer Brown's lambs had just been sent to slaughter as the courgette glut was ramping up in the patch adjacent to their field. So when the lambs returned, ready to eat, it seemed right to combine the two.

A certain solemnity pervades the departure of the

sheep from the farm and there's more than a little
soul-searching when I come to cook the lamb, as
I remember them watching me while I planted the
courgettes in the spring. But they were reared with
love, killed humanely and I reckon that if I can eat
them with respect and care then I have done right by
the flock. Besides, it's a good thing to come face to face
with the moral complexities of meat eating sometimes.
It stops me getting complacent.

So, for supper that September we celebrated the bounty
of the farm: the gorgeous lambs which grew side by side
with the abundant courgettes.

SERVES 4

4 big courgettes
3 tbsp extra virgin olive oil
400g lamb mince (20% fat)
1 tsp za'atar
1 red onion, finely chopped
2 garlic cloves, crushed
1 tbsp tomato purée
1–1½ tsp rose harissa paste (depending on much heat
    you like)
6 dried apricots, chopped
3 medium tomatoes, chopped
100ml red wine
30g pine nuts

Pre-heat the oven to 190°C.

Cut the courgettes in half lengthways and scoop out the seeds from the middle to create 8 courgette boats. Drizzle with 2 tablespoons of the olive oil and season with salt and pepper. Put them on a baking tray and roast in the oven for 30 minutes.

Meanwhile, sprinkle the mince with the za'atar, season with salt and pepper, then fry in a very hot, dry frying pan until browned and crispy. Remove to a bowl with a slotted spoon and set aside.

Turn the heat down low, add the remaining olive oil, then sweat the onion for 10 minutes, adding the garlic, tomato purée and a pinch of salt in the final couple of minutes.

Return the mince to the pan together with the harissa, apricots, tomatoes and the red wine. Allow the mixture to bubble for 10–15 minutes so it reduces to a thick, glossy sauce. Check the seasoning.

After their 30 minutes, remove the courgettes from the oven. Stuff each cavity with the mince mixture and top with pine nuts. Return to the oven for a further 15 minutes, then serve immediately.

October

I was, slowly, reintegrating

Simplicity

IN THE VEGETABLE PATCH...

October is special. It has its own vegetable, which makes it unique among months. Where December's logo is a Christmas tree, July's crest a tennis ball, January's coat of arms a snow-drop, only October has a vegetable. The pumpkin.

You might expect me to be delighted that a vegetable should take centre stage in this way. And I suppose I am. Celebration of any vegetable is a good thing but fame comes at a price. Millions of tasteless, watery Jack O' Lantern pumpkins, with their lurid orange skins and bloated middles, are grown for sacrifice beneath the carving knife at Halloween. It makes me sad. For two reasons.

First, there are hundreds – literally hundreds – of winter squash varieties, most of which are far more beautiful and strange than the carving varieties. They can be green, blue, orange, yellow or tiger-striped; smooth, knobbly, round, long, trumpet-shaped, warty or baboon-bottomed. They have weird and wonderful names like 'Speckled Hound', 'Golden Hubbard', 'Turk's Turban' and, one of my favourites, 'Sweet Dumpling'. Put them together and it all looks a little bit made up, the fevered creations of a surrealist sculptress – garish, bizarre, a little grotesque. Perfectly ghoulish enough for the season. But they don't get a look in. They live in the shadow of the dense, orange monster dominating the field like an American president and are mostly ignored by the majority of pumpkin buyers.

Secondly, what a waste! Jack O' Lanterns might not be my favourite squash (they don't have nearly the complexity of flavour offered by other pumpkins), but I still have a grower's aversion to food being wasted. It might be a thug, but it has feelings. How would you like it if you'd wished your whole life to be lovingly fanned with sage and roasted, or at the very least made into soup, but instead ended up being butchered by a cack-handed toddler and left to rot on a doorstep?

I grow a different pumpkin variety every year. They ramble around the bare soil beneath my sweetcorn or make their way up wigwams of poles that I always imagine are sturdy enough to take their weight, but never are. I pick them late in the month and store them in the tool shed over winter. It's a race to see who gets to eat them first – me or the shed-dwelling mice.

Crispy Parmesan Squash with Salsa Verde Butter Beans

To make supper for two, cut a small winter squash (an onion, kuri or Turk's Turban squash, weighing roughly 500g, is the perfect size, but half a butternut squash will do the job too) into wedges and de-seed. Leave the skin on, assuming it's not irretrievably gnarly. Douse in a slug of olive oil and season with salt and pepper. Roast on a baking tray in a hot oven (200°C) for 30 minutes or until just starting to char.

Grate a handful (50g) of Parmesan into a shallow bowl and, with asbestos hands, toss the wedges in the cheese, pressing them down so the cheese sticks and coats the flesh. Turn the oven up to 210°C. Return the wedges to the baking tray and roast for another 15 minutes, by which time the cheese will be crispy and irresistible.

While the squash cooks, heat a jar (not tin) of cooked butter beans (roughly 400g drained weight) in a saucepan with a tablespoon of extra virgin olive oil. Once warm, stir in a tablespoon each of chopped parsley, mint, basil and chives, 4 chopped anchovies, a teaspoon of baby capers, a shake of garlic granules (retro, but they are just the ticket here, I promise) and, just before serving, a slug of red wine vinegar.

Tip the butter beans onto a platter, pile the crusty squash on top and arm your fellow eater with a hunk of bread for scooping.

There is a stately home in Surrey which was once grand but has fallen on such hard times that it has been sold and turned into a golf course and luxury spa hotel. When you stay there, the waistcoated staff bring you the *Daily Telegraph* with your breakfast whether you request it or not ('I am not aware of another newspaper, Madam'). At dinner, they give the priced menu and the wine list to the men at the table, and unpriced menus to the women, who are apparently of too delicate constitutions to endure the vulgarity of the eye-watering prices and far too respectable to drink. Bolted on to the main building, like an architectural leech, is a bland modern extension that houses most of the rooms and could, were it not so well supplied with cushions and pelmets, pass as a low-security prison. At the entrance to the main building, a Cedar of Lebanon tree, which knew the house in its prime and remembers the wicker loungers and croquet on the lawn, surveys the whole monstrosity and droops symbolically, forlorn and round-shouldered in defeat.

In fact, there are many deflated mansion-turned-hotels around the South of England because, when not hosting golf clubs, they welcome corporate awaydays, and it is big business. I am trapped in such a place now. It is 2009 and I am nearing the end of a two-day workshop about margarine.

I am in my room, which is painted taupe and adorned with mushroom-coloured soft furnishings, with a fawn bathroom. I have stayed in so many of these hotel rooms that they have become interchangeable, and I know I will wake up tonight not knowing which one I am sleeping in. I change out of my Take-Me-Seriously black jacket and into a Look-I'm-Relaxing-Now-(but-not-off-duty) dress for dinner with the clients, where I will make polite conversation with the senior client about their family, their home country and what a beautiful building this is, yes, England is just like Harry Potter. Oh, your kids love Harry Potter do they? And how old are they? I am sure the client is as interested in answering these questions as I am in asking them, but we will play the game of small-talk nonetheless.

At this workshop we have had the Chief Marketing Officers (CMOs) from every major country of my client's business. People with corner offices and more than one secretary have travelled (first class, obviously) from Delhi, Singapore, Dubai, New York and Slough (the UK offices are never anywhere exciting). Research agencies have sent their Managing Directors to deliver their findings. We have our CEO in tow. I have had our design studio tarting up the PowerPoint presentation for days (this one is beyond even my PowerPoint art-direction skills). It is an expensive get-together.

We have one question to answer. It is a big question, and people's jobs depend on the answer. A lot of agencies' fees depend on it too, including ours. Hundreds of thousands of dollars have been spent researching the answer. We are told it could mean the success or failure of the brand; whether it is sold off and factories closed or whether it becomes a key pillar in the company portfolio. The stakes are high. The atmosphere is tense. It must be a pretty serious question, mustn't it? What with all this expense and all these smart minds.

Here is the question:

What is the purpose of this margarine?

Tellingly, it never occurred to me at the time that the only answer was, 'to spread on toast'. Because I had drunk the Kool-Aid too. But we didn't mean the *literal* purpose. We meant the reason to *be*, the mission, the philosophy, the beliefs. *Of a margarine.* Because people don't buy the *what*, they buy the *why*. They don't buy the 'spreads on toast', they buy the 'makes me a good mum'. Apparently.

I had pored over the research, I knew the insights, I found nuance and complexity in all the possible answers. I had just spent two days facilitating a workshop about it, dancing on the head of a pin in front of all these CMOs. Because my job too, as much as anyone's, rested on answering this question well.

That none of us thought to wonder whether this fuss was strictly necessary, or, more existentially, whether a margarine could really believe in anything anyway, makes me see how entangled we all were. In fact, it's only recently that I have come to see the full absurdity of it. But the processes were such that millions of pounds in sales, and countless jobs, rested on the decision, so the workshop was very heated.

Each CMO there represented a different country where

consumers had their own motivations, supermarkets their own demands, and cultures their own sensibilities. There was no hope of finding a single brand point of view which would satisfy all these differing factors. No way to accommodate the countless nuances of each market in one solution. Plus, CMOs can get, understandably, quite possessive about their country being 'different' and are prone to tetchiness if you prioritize one market over theirs, even if that market brings in five times the sales. And how they go about saying this to each other can be quite culturally, well, unique. The Italians do not disagree in the same way as the Chinese. Managing these sensitivities, bringing everyone into alignment and juggling all of those variables would have tested even UN peacekeepers. But I guess the peacekeepers were busy doing, you know, *important* stuff.

I come home from that workshop on a Friday afternoon, change into my muddy jeans and head straight into the garden and my new little vegetable patch. At the moment it is no more than a square of soil I have marked out with twine and tent pegs. We have only just left London and the raised beds are yet to be built (then neglected).

I sow a row of radishes. Having achieved so little in the past two days despite such effort, I have an urge to create something straightforward. I make a shallow line in the soil, sprinkle the seeds in, cover them up and water. No choices, no decisions, no questioning the greater purpose of the radishes. Just a simple act and something achieved. I wouldn't have described this as a medicine at the time, but it was. An antidote to the frustrations of the working week.

The radishes would be ready in just four or five weeks. The question of the margarine's purpose would still rage on as I picked those radishes. I would take them back to the kitchen, brush off the soil, dunk them – ironically – into soft, rich butter and sprinkle them with flaky salt. I would pop the whole lot in my mouth, crunching and waiting for the salty fat to give way to the peppery kick of radish. A whole plant would be sown, grown and eaten, an entire life cycle in nature would occur in less time than it would take us to not decide what to do with a margarine brand.

The act of planting those radishes was instantly calming. The eating too was simple and satisfying. I should have realized then that the vegetables would come to the rescue a few years later.

A Sauce for Radishes

Tip a 100g jar of anchovies in oil (or two 50g tins), into a food processor, oil and all, with 40ml olive oil, 1 fat garlic clove, a teaspoon of Dijon mustard, 2 teaspoons of apple cider vinegar, a pinch of chilli flakes and 5–6 tablespoons of hot water. Whizz to a smooth sauce. Check the balance of flavours – it will pack a punch – and adjust as needed. Left for long the sauce may separate, but can be whisked back together. Serve a bunch of radishes, sitting on a bed of ice for extra theatre, with a small bowl of the sauce beside them for dunking.

Never give a depressed person too many choices. It's paralyzing. I'm not saying choice is a bad thing per se, on many levels it is the privilege of liberty. But there are so many choices. And they have to be made *all the time.*

My job, like most, was a litany of decisions. And the more senior I got, the more decisions there were. What shall we do with this margarine? For example. These decisions aren't generally matters of life and death, not very major decisions, but their importance can get blown out of proportion in the pressure cooker of the office. The expectation that you will have an answer is intense.

Outside work there are choices too. There's the philosophical stuff, which tends to hit you if you are having a rough time: what am I going to do with my life? Where shall we live? Will I dye my hair? But the small decisions we make every day can stretch to the thousands too. Before you even get to work you have wondered: what shall I wear? Tube or bus? Do I want venti or grande? Soya, almond, oat or full fat? One shot, extra shot, dry, skinny, decaf, flat? Chocolate, no chocolate? Eat in, take out?

The croissant I want or the coconut yoghurt I ought to have?

As the number of options on offer to us has grown, so the number of micro-decisions before us has increased exponentially over the years. Take chocolate bars (I worked on a lot of chocolate brands). Where there used to be one format of your favourite chocolate bar, there are now dozens – duo bar, fun-sized, sharing bag, large block, dessert pots, with marshmallows, without marshmallows, extra protein, more nuts, fewer nuts, limited edition, soft-melt round-edged easy-access single-serve re-sealable. And that's before you even get to the social decisions about your chocolate bar: have I deserved it? Should I feel guilty about eating it? Is the cocoa ethically sourced? Can I recycle the packaging? What does the brand say about me? What will people think of me when they see me eating it? Will it reflect on my morals as a human?

This is my doing. It was my job to add to these questions because they affect people's purchasing choices. I built a career out of making the decision as loaded as possible.

For the record, I do not think life was better in the olden days when choice was limited and things were simple. Life might have been simpler then, but simple isn't always better. Women, for example, had simpler lives, but that was because they couldn't have a career, vote, control their own money or own property. So choice is progress and freedom, undoubtedly.

But. You need your wits about you to deal with it. Our world is so full of choice that, when I lost my wits, I found myself overwhelmed by it. I could not even handle the small stuff, let alone the job-critical choices about margarine brand essence and chocolate formats. I was so overwhelmed by the barrage of endless small choices that I couldn't even contemplate the big choices. If I couldn't decide whether I wanted a cup of tea, how could I even imagine deciding to quit my job?

In Ad-land, where pseudo-psychologists are ten a penny, we called this 'choice paralysis': the panic that bubbles up when you have too many options to choose from and end up avoiding the decision all together rather than attempt the Herculean task of weighing up the options. When I was depressed, I would get choice paralysis with *everything*. The micro-decisions that are occasionally slightly wearing when you are well become towering monoliths to hurdle when you are ill. Deciding what to have on my toast, what to wear, when to shower became panic-inducing challenges. There were days when I couldn't hang the laundry up because the effort of having to decide where to hang stuff was too great a decision-making challenge. Should all the pants go together, or can they mix with the socks? Do shirts need to go on hangers or on the rack? Had that jumper better dry flat? To these questions, my mind replied, 'Christ, I don't know. I don't know. What am I going to do? Cannot compute. Panic stations.' It was just doing the laundry, but I couldn't do it.

In the vegetable patch, I felt none of this. You didn't have to make any decisions in the patch because it all ultimately came down to some very simple, immovable rules: water stuff, it grows. Don't, it dies. When you do what you are supposed to do, stuff goes well. When you don't, it does not. Yes, nature intervenes sometimes with an unexpected pest that eats your crop or a random hot spell that fries everything. But that's not in your control. Not for you to make a decision about. Above your pay grade. It just happens. Goal posts do not move in the kitchen garden. There is an order to follow and all you have to do is plod along doing it.

Take carrots. You sow carrots thinly, in sandy soil, anytime from April. You water them. And wait. Yes, there are choices about what variety and where to sow them, but to me these felt like creative decisions that were just a matter of preference rather than a predictor of success. No one would reflect on my carrot-growing capacity; the approved method for carrot growing would not change in the time it took them to grow; no CMO would come and pronounce my carrot harvest the wrong colour for her market's needs. I was clear about what nature wanted of me, and if I did that, my carrot harvest would succeed. It felt obvious in the garden because I understood how things worked here in a way I never did in Ad-land. Here there were no unexpected additional criteria that appeared from left field half way through carrot growing to scupper your efforts. With my hands in the soil, things were simple; I understood them in a way I never did in the prison-wing extension of a Surrey hotel.

Why did I feel I understood nature in this instinctive way? It wasn't as if I grew up among rolling fields, collecting worms and making dens. I lived in the suburbs. I'd spent more time in the town shopping precinct than in the countryside. We didn't even go on camping holidays. It was not at all *Swallows and Amazons*.

Only in researching this book have I even begun to answer this question. I have read about 'biophilia' (thanks to Johann Hari's book *Lost Connections*) – an idea, popularized by the biologist Edward O. Wilson in the 1980s, which describes the human drive to connect with nature and argues that it is genetically predetermined, an innate need. Wilson writes that spending time in nature reduces our blood pressure, depresses cortisol levels and restores our attention levels, which is why it makes us feel good.

As Hari says, that makes sense. We are animals. Animals who have existed in nature for a thousand times longer than they have lived in office blocks. No wonder I felt like I was home. I *was* home. I might not have grown up in nature, I might not have a horticulturalist's expertise, but at an evolutionary level this was my natural habitat.

And so, I instinctively turned to the garden at my lowest ebb. With this regular dose of what was real, truly real, I found I could function in the 'constructed' world of modern life better. That month, for example, I took the train to London for lunch with old work friends from a previous agency who knew my situation and so met me at Paddington. I went to the cinema to watch the new James Bond and didn't want to hide under the seat because of the noise. Both excursions felt like I was on day-release, my world having become so small, and the logistics of booking tickets and so on made me fret. Paul checked the bookings before I paid online and punched in my card details because I couldn't get them in order. But I could now manage perhaps two or three 'outside' activities a week without getting overwhelmed – just as long as I made sure there was time to get my medicinal shot of veg patch in between where things remained calm and simple.

Everything felt simple in the veg patch because I understood it. I understood it because it was familiar. It was familiar because it was nature. And nature was home. That's why, after my margarine workshop, I turned to a row of radishes for comfort. Now, again, the veg patch was, delicately, and with the wise forbearance of a good psychiatrist, offering a simple, safe environment that helped me handle the rest of the more cluttered world. I was, slowly, reintegrating.

Radish, Cucumber and Whipped Feta

The nice thing about radishes is that they grow quickly. Five weeks, with warm days and a fair wind, and you will have a crop of little red jewels – which is less time, as I've said, than a bunch of smart people can spend failing to agree on the *raison d'être* of a margarine.

This recipe is a calming and simple affair. I like it because it is so pure, so unadulterated that you can concentrate on the peppery kick of the radish, the clean greenness of the cucumber, undisturbed by fussy dressings or adornment. It is so gentle and self-contained that I find it almost meditative. These are vegetables so quietly confident in their beauty that they feel no need for embellishment; a total contrast to the flashy, grotesque dishes that filled the tables at advertising lunches. I would take this salad, made with home-grown vegetables, over an aerated lobster bisque and seaweed reduction any day, even if it came with a tuile.

SERVES 2

8–10 radishes
½ cucumber
1 tsp chopped chives
1 tbsp extra virgin olive oil
Juice and zest of ½ lemon
200g feta
1 tbsp chopped parsley
2 tbsp crème fraîche

Use a potato peeler or mandoline to slice the radishes into very thin, round slivers. Take the peeler to the cucumber too to make a pile of cucumber ribbons. Toss them together in a bowl with a pinch of good-quality flaky salt and the chives, olive oil and lemon juice.

Crumble the feta into a herb chopper or blender, add

the lemon zest, parsley and crème fraîche and whizz to
a creamy, mint green dip. Spoon the whipped feta onto
a plate and pile the radish and cucumber on top. Finish
with a twist of pepper and a dash more olive oil. That's all.

November

Nature would hand
me back the reins

NOVEMBER

Self-sufficiency

IN THE VEGETABLE PATCH...

November is, for me, like the morning after a great party. You had a brilliant night, but now it's time to drag a black bin-liner around the living room collecting empty bottles and nursing a hangover. The summer harvests are finished, the squash are in the shed and the sweetcorn are plucked. All that's left of the summer fun is a tangle of bean canes and the knot of squash vines, brown and crisp. It's time to clear up. As you gather up the slimy, frostbitten courgette plants you might discover a forgotten kohlrabi underneath a leaf, like finding a sleeping partygoer from the night before under a pile of left-behind coats.

Some growers find this part of gardening satisfying. They like to create order out of disorder and beaver away at the tidying like a proud 1950s housewife. I am not that grower. To me it's just housework, but outside.

It's not all drudgery though. The brassica patch, defended all summer from cabbage white caterpillars by mesh netting and a rickety arrangement of bamboo canes, is home to cabbages, kale and cavolo nero, all ready for picking in November. Elsewhere, a few carrots and beetroot remain in the ground. They won't grow any more, but the soil acts as larder storage for fully sized roots. But beware, because, unlike leeks and parsnips which relish a frost, if carrots and beetroot get frozen, either by a ground frost or by touching the back of the fridge, the results will be the same – mushy roots. Still, keep an eye on things and you will be rewarded with something to eat. A little taste of self-sufficiency.

As a child, I loved watching *The Good Life*. It was a seventies sitcom about a suburban couple, Tom and Barbara Good, played by Richard Briers and Felicity Kendal, who quit their dull office jobs to become self-sufficient. They bought a goat (Geraldine) for their semi-detached back garden, turned the front garden into an allotment, that sort of thing. Needless to say, the keeping-up-with-the-Joneses neighbours in the cul-de-sac were scandalized and hilarity ensued.

Mainly I liked Felicity Kendal's outfits, and how they were grown-ups but still made things out of old newspaper and cereal boxes like I did. But also, and I could not have explained it like this when I was seven, what I think I really liked was the way they made their own happiness. Happiness wasn't dependent on external factors like money and possessions as it was for the Leadbetters (their neighbours); Christmas wasn't cancelled when the department store did not deliver all the accoutrements they needed for it (as happened to poor Margot Leadbetter in a Christmas special episode). Tom and Barbara were self-sufficient in more ways than just their food. They had agency and control, and their happiness was entirely their own making. That emotional and physical self-reliance appealed to me.

Depression robs you of any sense of self-reliance. I felt utterly helpless. But then I think I had felt that way beforehand too. I was so busy with work that I outsourced everything to do with my personal life. I had clothes cleaned, the ironing collected, food delivered, travel booked, the lawn mown. If there had been a dial-a-washer-upper app I would have used it. My food was not only grown by someone else, but chosen, cooked and delivered too. I could afford all this because my job paid well. But, in order to keep that job I had to dedicate all my time to it, which meant outsourcing anything else that took up time. It's a common trap: work hard to make enough money to pay to free up time to work harder. The result was that I did nothing for myself. I was nothing like Tom and Barbara.

In his book *Cooked*, Michael Pollan talks about the division of labour in modern life and suggests that, 'before long it becomes hard to imagine doing much of anything for ourselves'. Division of labour is generally a good thing, he says. It means that he, and I, can sit here writing words at a computer without having to mill some flour to make bread for our lunchtime sandwiches, or milk the cow to get milk for our bottomless mugs of tea. But take it too far and we can develop a learnt helplessness.

And if you feel a little helpless *now*, if you think you have ceded control *already*, wait until you add depression into the mix. With depression, your mind is no longer under your control anymore either. You forget things. You make stuff up. You cannot compute what happens around you. If you had asked me that November what month it was, I wouldn't have known it, I would have had to deduce it.

'Well,' I would say, 'the evenings are dark, we had a fire last night and apparently it's my birthday soon, so I'll say it's November.' Bingo.

Whole chunks of that autumn remain a blank. In writing this, I have had to corroborate things with my family. I thought, for example, that Mum had come to sit with me during the day-time for a week or two. It was three months. I still cannot place events or conversations in order from that time. My memory was not my own.

Sleeping pills did nothing but exacerbate this. It is my pet theory (based on nothing more than anecdotal research) that if you cannot get a decent night's sleep, you are lost. Jet lag, teething babies, money worries, grief, whatever, can all descend into mental illness if they mean you do not sleep. So I can understand why doctors prescribe sleeping pills: get the sleep sorted and that will help fix everything else.

However, sleeping pills come with side-effects (which are highly individual and some people have none at all). And for me these side-effects outweighed any benefit. Yes, you will technically be asleep, but you will fall asleep feeling paralyzed, the energy in your body still ricocheting around your limbs, which are restless and desperate to move but somehow won't respond to your command. And when you do finally sleep, you are visited by nightmares of such violence and horror that you will question your sanity, what's left of it, when you recall them. What kind of a brain have I got in there that could imagine such things? In the morning, you will wake up groggy, hungover and exhausted from your dreams, the trauma of which will feel very real and will linger all day. You will have no sense of time having passed either, and you spend your first confused moments wondering how it can be light when you shut your eyes only a second ago. I have never had an anaesthetic, but I'm told the sensation is similar.

On one such morning, my mum took me to the Ashmolean museum in Oxford. It was a valiant attempt to get me out the house and a good choice: calm, not too busy, nothing too loud; stimulating but not taxing. Well done, Mum. But I had taken a sleeping pill the night before and I was narcoleptically sleepy. I could hardly stand up. My eyes couldn't focus. My speech was slurred. I sat on one of those padded benches they put in front of the famous statues so you can ponder their beauty at your leisure. My head lolled, my mouth gaped open and I slumped forward.

People stared piteously at my mother. 'What a shame to have such a burden', their looks seemed to say.

A nervous teacher ushered her gaggle of impressionable ducklings away from the strange, spaced-out madwoman.

'Let's go and draw the sarcophagus instead shall we, children?'

No, sleeping pills did not help me. They just took away even more control.

Homemade Horlicks for a Restful Night

I had read somewhere that a hot, sweet drink before bed would help you sleep, so I cultivated a Horlicks habit, taking one to bed each night in the hope of sleep. I'm not sure if it helped, but it was certainly comforting. The latest science, though, would suggest that the worst thing you can do for your nighttime glucose levels and the health of your microbiome is to have something sweet late at night, so here's a slightly less processed, marginally less sweet alternative.

In a small saucepan, whisk 200ml of oat milk (regular milk will work fine too, but oat is sweeter and has a comforting toasted-cereal taste already redolent of Horlicks) with a teaspoon of malt extract, a teaspoon of cocoa powder and, for extra interest but not strictly necessary, two crushed cardamon pods and a stick of cinnamon (which I reuse for a few brews). Bring to a gentle simmer for a couple of minutes then pour into a mug and climb into bed.

That autumn, growing vegetables gave me back some control, some small notion that I could be self-reliant again. When I was well, I had everything done for me. When I got sick, even more was ceded. By November, while I could, for example, sit through a whole film now, without the fizzing in my head making me fidget too much to sit still or a character's heartache sending me into hysterical tears (which was progress), I had no ability to decide that I was going to watch something or what I wanted to watch. Instead, my dad, a film buff with drawer upon drawer of DVDs, would send a DVD with Mum on her daily visits and I would be plonked in front of the TV where I watched it unquestioningly. It was such a small choice, the decision to watch a film and the picking of it, and it should be a pleasure, but I needed someone else to make those decisions for me and was happy to be sat passively in front of the TV like a child.

Nature would hand me back the reins. I went up to the patch with an empty trug and no supper and returned with a harvest of kale, cabbage, beetroot and borlotti beans, my mind buzzing with ideas of what to feed myself. I had created the harvest and would create the meal. I had instigated it all. I wonder now how a brain that couldn't muster the cognitive capacity to pick a movie could tend a garden, but I think the difference was that in the patch a process was ordained by nature. She had already decided, and, if I showed up and just did the work, things would progress. But *I* had to do it.

And for me, this is the single, most important thing about how growing your own food affects mental health. This is where it differs from simply gardening or getting out in the countryside. *Agency.*

Let me explain. When you grow something to eat, you have taken back control of your existence. You have, single-handedly, without any help, without having to be dependent on other people, made a meal for yourself. You have done one of the most essential things required for life – fed yourself. And in doing so you have seen that you are capable, that you can look after yourself, that you are in control of the most fundamental parts of your life (if not movie choices).

Pollan puts it neatly: 'taking back the production and preparation of even just some part of our food is to take back a measure of responsibility ... to undo our learned helplessness'.

In other words, we become more like Tom and Barbara Good. Being the agent of change, being the one who had a hand in turning a seed into your own supper, gives you a profound sense of independence.

This is why growing vegetables is not 'just' getting out into nature. The effect of growing food on your well-being is fundamentally different from other forms of gardening or ways of connecting with nature. It makes you feel *empowered*. It makes you independent. Growing some flowers or going for a walk in the hills might reconnect you with the cosmos, but only food-growing will give you power.

Not only are you more self-reliant on a personal level, but

you are also dismantling the systems that keep us dependent. If you can feed yourself, you will be a little less reliant on practices that harm the planet and exploit growers. You are a little less at the mercy of advertisers (advertisers just like I had been) who make you want things you didn't know you needed and probably aren't very good for you. Even if it's only on a tiny scale, being able to grow some food allows you to opt out of that system, just a little bit. Growing vegetables is a tiny piece of defiant self-sufficiency. A quiet act of resistance.

When I think about this, I am always reminded of *WALL-E*, a Pixar movie set in the future where the humans have 'evolved' into helpless, obese babies who are fed meal-replacement shakes by robots and ferried around on hoverboards, passively consuming one leisure activity after another (such as watching whatever movies are fed to them – sounds familiar) like they're on a warped cruise holiday. This is the dystopian world that growing your own food, reclaiming some agency in your existence, will save you from. It allows us to imagine a different future.

Vegetables. You didn't think they were so political, did you?

Empowering Kale Pesto

Preserving a glut makes me feel like a prepper, in the best way. I batch cook something basic, freeze it or jar it, then store it away for when the revolution comes, and I can still feed myself.

Kale pesto is perfect for this. Simply stuff as much kale as you can into a food processor, add some grated Parmesan (it needs to be grated even though you're blitzing it – I tried cubing it once and it was like eating gravel), a couple of garlic cloves, a handful of nuts (hazelnuts, walnuts, pine nuts, or even pumpkin seeds), a big slug of olive oil and a little salt. Whizz. Adjust the flavours and consistency as you see fit. Spoon into small pots. Freeze. Then whip out a pot as needed to stir into hot pasta, warm butter beans or to spread over fish, chicken, or whatever is available once disaster strikes. So much more satisfying than a cellar full of tins. Just so long as the power stays on…

But let's not get giddy. I'm not suggesting we all sell up and go off-grid to improve our mental health. I am not enough of an impressionable romantic to imagine I would last more than five minutes if I tried to live self-sufficiently. I would find the whole thing hugely stressful and overwhelming. But total self-sufficiency is not necessary. If the goal is to remind ourselves we have more self-reliance than we might think, then even some small act – like growing radishes – will be enough to have an effect.

And it did for me that autumn. Emotionally I was still fragile, I was still dependent on others for many things, but I was beginning to see that I could take ownership of small aspects of my life – most significantly, what to eat.

Self-sufficiency Stew

Though I am far from being self-sufficient, and I don't look nearly as good in dungarees as Felicity Kendal does, I do like to try and make a meal in autumn that is as home-grown as possible. That's easy to do in the summer, but it requires a bit more of Barbara's ingenuity by November, when the harvests are more sparse.

This stew is an autumn ritual and is made with the borlotti beans I should harvest in late October but usually forget until November, and whatever else I can extract from the patch.

SERVES 2

75g pancetta
1 onion, finely chopped
1 celery stick, finely chopped
1 carrot, finely chopped
2 baby leeks (if they are ready to harvest),
    finely chopped

500ml chicken stock
250g fresh (i.e. not dried) borlotti beans,
   or 1 x 400g tin, drained
150g savoy cabbage, finely shredded
100g small cavolo nero leaves
A few kale leaves, stalks removed and shredded
2 garlic cloves
2 tbsp chopped parsley
½ lemon, zested
1 tbsp extra virgin olive oil

Fry the pancetta in a casserole dish over a high heat until golden brown. Turn the heat down and add the onion, celery, carrot and leek, and a pinch of salt. If the pancetta has not released much fat, you might need to add a slug of olive oil to the pan. Sweat the vegetables for 10–15 minutes until they are soft and translucent.

Add the stock along with the borlotti beans, cabbage, cavolo nero and kale and bring to a gentle simmer. Leave to bubble, with a lid on, for 5–10 minutes or until the beans are soft and the leaves wilted. (Fresh beans will take a few minutes longer to soften than tinned.) Check the seasoning and adjust as you go.

In a pestle and mortar, crush the garlic and parsley together with a pinch of salt, then muddle in the lemon zest and olive oil to make a stiff paste. Do not omit this bit. It's a gamechanger.

To serve, ladle the stew into two bowls and top with the garlicky paste. Some crusty bread wouldn't go amiss here either.

December

lifting me out of
melancholy

DECEMBER

Nurture

IN THE VEGETABLE PATCH...

The first year I grew vegetables, I tried growing Brussels sprouts for Christmas. They failed. I tried again the next year, and the one after that. They all failed. I love sprouts. I love Christmas. I love that sprouts and Christmas happen at the same time, but I'd settle for sprouts any time of year. If only I could grow them. But I cannot grow sprouts.

Brussels sprouts are quite technical. To be frank, they are tricky little blighters. For a start, they take almost a year to grow. Sown in a seed tray in early spring, they are transplanted into individual pots after six weeks, then into their final positions in June to grow on until harvesting in December at the earliest. They like their own space and expect a solid 60–70cm between them and the next sprout tree. They are fussy, too, about having their roots firmly secured in the ground and will 'blow' (i.e. the sprouts won't form tight, firm balls of leaves) if they are left to rock about in the wind or loose soil. So they need staking as well. They are at their sweetest after a frost, which presumably is why they became associated with Christmas dinner. If you get that far.

Much easier to cultivate for a December harvest is the Kalette or 'flower sprout'. This is a cross between (can you guess?) a kale and a sprout. They are more amenable to my haphazard growing style and less prone to wind rock and pests. Just like a sprout, they have fat trunks on which grow little baby cabbages. But the Kalette's mini cabbage is a loose-leafed floret of wrinkly leaves rather than a tight ball of smooth ones. They look like mini butterhead lettuces, plucked from the kitchen garden of a doll's house. Kalettes are sweeter than Brussels sprouts too, so do not induce the same face-wrinkling wince when you serve them up.

Once I eventually came to terms with my sprout-growing inadequacies, I turned to Kalettes instead and have grown them, quite successfully, ever since. We've eaten them around Christmas for so long now that I've come to think of them as an essential part of the festivities.

Kalette Noodle Salad with Savoury Seed Granola

For the granola, I must warn you: make more than you need – maybe double the amounts below – because it's so addictive there is a high attrition rate between making and serving up. Also, it keeps in a sealed container for a few days. If it lasts that long. Use any noodles you prefer, or none at all – the main point is the Kalettes.

SERVES 2

75g edamame soybean noodles (I use Yutaka)
    or any noodles you fancy
170g Kalettes
1 avocado, sliced
1 apple, cored and sliced
2 tbsp dried blueberries
1 tbsp sesame oil
2 tsp mirin
½ lime, juiced

FOR THE SPICY SEED GRANOLA:
1 tbsp sunflower seeds
1 tbsp pumpkin seeds
½ tbsp chia seeds
½ tbsp sesame seeds
1 tsp soy sauce
1 tsp honey
1 tsp Worcestershire sauce
Splash (or two) of Tabasco
1 tbsp olive oil

Start with the granola. Mix everything together in a bowl, then tip into a dry, non-stick frying pan and fry gently for 3–4 minutes, stirring continuously, until the seeds turn a rich, dark brown. Beware: this can happen suddenly so keep an eye on it. Tip onto a silicone sheet (or a tray lined with non-stick baking parchment) and leave to cool. The granola will crisp up, set and clump together as it goes colds. Keep in an airtight container until needed (it will stay crunchy for a couple of days).

For the salad, cook your chosen noodles as the pack instructs, then rinse under cold water and set aside.

Steam the Kalettes for the briefest of time (2 minutes max), then plunge into cold water to keep them from over-cooking, pat dry gently and pop in a bowl.

Add the avocado, apple, blueberries and the noodles. Then pour in the sesame oil, mirin and lime juice plus a pinch of salt. Toss everything together gently and tip onto a plate. Scatter over the seed granola and serve.

It would not be accurate to give the vegetables all the credit for reinstating my sense of purpose that autumn. Sure, the acupuncture, weekly therapy sessions and antidepressants continued. So, too, did the accommodating, careful support of my bosses, who continued to put their hands in their pockets, encouraged my allotment as a creative outlet, listened politely when I said I'd be back in the saddle by January, and hinted that a day a week at most might be wise. But more significantly than all of this, around this time, we got a dog.

We had always wanted a dog. We decided on the name years ago – Hadleigh – and we knew we wanted a working dog like a German pointer or an English springer spaniel. But working in London and travelling had made it impossible. Now I was at home those barriers were lifted. Getting a dog had not occurred to me, but getting dressed didn't occur to me then either. It was Paul who suggested it one day and I agreed immediately. It is the only time that having no sense or critical faculties has been useful, because if I had properly considered it, we would never have done it.

Paul thought a male springer spaniel would be best and I agreed. Usually, I am not so casual about decisions like this. I like to look at all the options and develop my own, quite strongly held, opinions. But I was in no state to think and so I just agreed and said I would start looking for a suitable breeder.

I found a breeder near Bristol with two five-month-old puppies left from a working litter. The breeder didn't tell me at the time, but it turned out that one was the runt of the litter, and the other was so naughty no one would take him.

My mum drove me to Bristol to have a look at these, unbeknown to us, rejects. We got lost because I couldn't fathom the directions or work the maps app on my phone.

I got in a flap about it and Mum said, 'Darling, given that you can't read a map at the moment, is it sensible to expect yourself to look after a dog?'

Ah, the voice of reason. But reason never entered anyone's head when they got a dog. There is never a good time to get a dog. So, when I fell in love with the runt because he came and sat on my feet and gazed at me, sensing I felt as dejected as he did, we went back the following weekend and got him anyway.

The first few weeks were hell. Hadleigh turned out to be eccentric, nervous and sensitive. He was wild: aghast at being put on a lead, terrified of everything, panicky about being indoors. On the first day we had him he didn't make a sound all day. I thought he was mute. But when we tried to shut him downstairs at bedtime he went crazy, barking and throwing himself at the door in wild terror. He had never lived in a house and was afraid of being contained. He kept walking into glass doors, puzzled at being able to see the outside but not reach it. When we tried to introduce him to upstairs, he just looked at the stairs, then looked at us as if to say, 'What in God's name do you expect me to do on those?'

But all new puppies are hard work, and these were quite endearing difficulties. It was my reaction to Hadleigh that was the real problem. I thought I was failing every time he barked. The noise made me shake. I was so nervous I couldn't inhale when we first let him off the lead. I was constantly terrified he might run away, or bite someone. I was paranoid about him weeing in the house so I took him outside obsessively. Every tiny misdemeanour was escalated in my misfiring mind to crisis level, a reflection of my failure and just another illustration of my uselessness. After one particularly barky night, I was so delusional I had convinced myself that Hadleigh just hated being with me, that he could sense my madness and that the best solution was to take him to my neighbour's in the morning (the one who was still diligently and unwaveringly messaging about tea) and give him to her, even though she worked full time, hadn't offered and had no intention of getting a dog. But in my mind, this was a perfectly feasible solution to a very tangible problem that the dog was allergic to my madness.

With Hadleigh's arrival came the challenge of dog walks.

First, getting out of bed. Some days, the effort was just too much. What do I mean by 'too much'? I think this – lethargy – is one of the hardest things to empathize with if you've never had depression. Feelings of grief or sorrow or agitation are something most people can understand. But the feeling that you just can't find the will or the cognitive function to stand up is unfathomable to many. It isn't like having flu, when you can't move because your limbs ache. And it isn't like being tired from lack of sleep or exercise when your limbs feel depleted. It is partly cognitive: I can no longer find the part of my brain where I make my legs move, someone has switched it off. It's partly emotional: what's the point of even trying? I'll just get it wrong. And it's partly ownership: these limbs, this body isn't really me anyway. I am stuck in my head, I am nothing but my addled mind and this body isn't under my jurisdiction any more. Plus, the duvet is warm and safe and perhaps I could just stay here until it all blows over. No, I can't get up.

Just say I made it out the front door, what if I saw someone? I was not fit to be seen. I was hideous. Strangers would recoil. It wasn't that I was ashamed of being depressed anymore. (Though I had been at the start.) It was more that it seemed unfair to subject others to the horror. A gibbering, grief-stricken mess who can't look you in the eye and flinches when you speak. It would be so awkward for them. No one needs to see that. Especially not looming before them on their dog walk like an unbidden ghost. They haven't even had breakfast yet.

But even worse than seeing a stranger would be seeing someone I knew. A stranger has no reference point so might be hoodwinked into thinking you are sane. But a neighbour might spot your shaky hands, your paleness, your weight loss. And what if they asked about work or why you aren't there? God forbid you should tell them the truth. Why ruin their day? Anyway, it's not the sort of conversation you have when you bump into a neighbour on a dog walk is it?

'Hi Kathy, how are you? New puppy? Isn't it a glorious day?'

'Dreadful. Yes, new puppy, do you want him? And no not glorious. Anyway, must dash, I have the whole day to spend fretting about whether you looked at me funny.'

Best not. Easier just to stay at home.

Things settled down. They always do. I put hours and hours into training Hadleigh and while the motivation at first was to assuage my guilt about being a terrible dog-owner and therefore a bad person, the training gave both of us structure and purpose. Eventually I saw results. The satisfaction I felt at having Hadleigh come when I called and sit looking up at me, was the same sense of achievement I felt when I harvested a perfectly formed cabbage.

'I made that happen,' I said to myself.

The following summer, when it took just five minutes to teach Hadleigh to walk around the vegetable beds at the farm to get to me, rather than take the more direct route over them, I looked back on this dark December and was glad of the effort I had put in to training him.

Hadleigh grew up to become a devoted, nervous and peculiar dog who assumed disaster was around every corner and always, *always*, did as he was told, however much it pained him, so as to avoid disaster as best he could. He died a decade later, before his time. Anyone who has lost a dog will know the very particular kind of pain it causes and the bittersweet heartache that comes with getting another dog, which we inevitably did. Through the grief, and with the arrival of Humphrey, also a springer spaniel, for whom the world is one big, joyful, welcoming adventure (quite a culture shock), I came to realize for the first time that Hadleigh had not simply been a comfort, or something to nurture; he had also been an ally when I most needed it. He and I were of the same mind, like a dæmon in Philip Pullman's *His Dark Materials* series: a physical manifestation of a character's inner self who takes animal form and accompanies them through life. We were both certain catastrophe was imminent, that something was bound to go wrong, but since we both agreed on that we felt a little less alone. It was only when I lost him that I realized he had been my dæmon.

In many ways, Hadleigh provided similar therapy to the vegetable patch. With both, I was connecting with the natural world. I was seeing nature develop and grow under my nurturing. And once they did grow, they were daily evidence of my agency, of my capability. So perhaps I wasn't that helpless after all. Perhaps I wasn't as alone as I imagined. I think there should be allotments on the NHS, but if I had my way, there would be dogs too.

By Christmas things are looking up. I have mastered the shower taps again. I can make tea – a landmark. Hadleigh sits when I ask him, and we discover we both have the same approach to life so now I have someone to share the anxieties with.

I have abandoned the sleeping pills and am sleeping unaided (though 'aid' is not how I would describe what the sleeping pills offered). A combination of antidepressants, CBT, meditation, acupuncture (turns out it even works for a cynic like me) and a summer in the vegetable patch has quelled the racket in my head.

How anyone gets over depression is very individual, if, indeed, anyone gets over it at all. Perhaps 'manages' is more accurate since, in my experience, it leaves traces long after the acute phase is dealt with. I am, for example, in two minds about the use of antidepressants. I was told to take them and was in no fit state to argue. I was also poorly enough that I would have taken anything I thought would rid me of the anguish, regardless of what else it took with it in the process. And the drugs did take away the strongest feelings of melancholy. But they also took any other strong feelings I had: love, interest, compassion, desire, pleasure, anger, motivation – the whole lot was switched off, and with it my humanity.

The CBT also helped me talk myself down from the worst of my ruminations and catastrophizing. But both were sticking plasters rather than solutions. They treated the symptoms not the cause.

For me, the non-medical therapies were what helped me rebuild a sense of identity. The drugs and therapy stopped the illness getting worse, but without the meditation, the acupuncture, the dog and the vegetable patch I would not have begun to address the causes of my distress. None of these remedies were suggested to me. I stumbled upon them myself, and I had the support, cash, time and infrastructure to access them. I was lucky. And I am not unaware that those without such access are far less-well cared for.

This combination of therapies, then, is working and I am now able to see, if not in full colour, then at least in halftones. Which is good because it is Christmas. And I am annoyingly, cloyingly, childishly excitable about Christmas.

I was sure that I could never be excited about anything ever again. I thought I had lost the capacity for enjoyment. But, at home with time on my hands, no weeding to do and some semblance of a functioning brain now, I throw myself into festive planning. And it is *fun*.

Food is the main feature of the plan. As long as no winter storms come to batter the vegetable patch, there will be Kalettes, red cabbage, parsnips, kale and celeriac for the table. I visit the patch almost daily to check on progress. The celeriac are petite, but I have planted far too many so, overall, volume will not be an issue, even if each root is only the size of a very muddy tennis ball.

A pigeon accidentally-on-purpose gets itself stuck inside the netting intended to protect the kale. Deciding to go out on a high, it gorges itself on the plants it is now imprisoned with, before collapsing in a gluttonous stupor and dying beneath the stump of a tattered, skeletal kale plant. The kale is rationed to ensure there is harvest enough for Christmas.

I have read something about how to have home-grown new potatoes for Christmas. Usually new potatoes, like Jersey Royals, are harvested in the summer. But I half remember reading that if you put this summer crop into a biscuit tin and bury it in the ground, then they keep for Christmas. But I buried the tin during a flaky episode back in August, so whether we have home-grown spuds on Christmas Day is dependent on me being able to remember where I put the biscuit tin. Christmas treasure buried in an unmarked spot.

(There is an easier way, I discover years later. Plant a fast-growing maincrop (old) potato variety in late August, then, once they have flowered, pile up the earth around them and fleece them until Christmas. As long as the ground isn't waterlogged, they will keep quite happily until Christmas and make much better roasties than new potatoes left in a tin would. Plus, you won't forget where they are.)

This first Christmas harvest was bounteous and has never been replicated in the years since. It was by far the best I've ever had, and I wonder if my newly kindled, bubbling desire to nurture was the reason.

At my lowest, I wouldn't have thought cooking was worth the bother. It was too much effort, and I didn't care anyway. I would eat if someone put something in front of me, but whether that was a gourmet plate of sushi or a fish finger didn't interest me; there was no joy in either. And if there was no one around, I mostly just ate toast. But my motivation was slowly returning and, that Christmas, it manifested itself in cooking.

The harvests from the patch were filling me with ideas. I wanted to braise my red cabbages, toss my savoys with bacon and nutty spelt, roast the first parsnips with miso, and mash the celeriac into salty butter to serve with venison. I stuffed a cabbage with Christmassy pork, sage and chestnut stuffing. I was full of recipes, inspired by my crops.

Paul was doing a valiant job of eating the endless dishes

I was cooking. He didn't even blanch when, returning from work one Tuesday evening, I announced: 'I've braised these pigs trotters for supper because I think they'll be perfect stuffed into this savoy cabbage.'

Braised Trotters and Cabbage

Well, since you ask, it's a matter of, in a casserole dish with the lid on, gently bubbling in the oven (at 150°C) four trotters submerged in a broth of water, one onion, two carrots, two sticks of celery, 30g dried mushrooms and a couple of bay leaves, for about 3 hours. Once drained and cooled, pick the morsels of meat from the feet. Finely chop this meat with, from the pot, the braised onion, one of the carrots, a celery stick and a few of the mushrooms. Season well. Stuff the minced mixture into four savoy cabbage leaves, rolling each parcel tightly. Brush with a little of the braising liquor and bake at 180°C for 20 minutes. Meanwhile, reduce the remaining braising liquor by at least 80 per cent, adding a splash of sweet sherry and some salt in the final few minutes to make a glossy sauce. Serve the stuffed cabbage leaves between two people with mash and the gravy. Easy...

But one person, however willing, could not afford me all the opportunities I needed for cooking my harvests. So, we had friends over. For the first time in months, we entertained.

To my surprise, I found I wanted nothing more than friends and family around the table eating food I had grown and cooked. I was so proud of my winter harvests and I wanted to show them off. A good pumpkin feels like a huge achievement at the best of times, but after months of incapacity it seemed like a miracle. Friends who had made me tea and toast when I couldn't get off the sofa came round and I made feasts of home-grown squash stuffed with cabbage, bacon and barley risotto.

I was still cautious. Only inviting a couple of people at a time so I wasn't overwhelmed. Never going to the supermarket for supplies at peak shopping times, since the busyness and the crowds were still panic-inducing. I was learning to manage my input levels and to pace myself.

The antidepressants had initially, in my case anyway, numbed all emotions, both positive and negative. But now I found that if I made a conscious effort to create an environment or experience that would bring me joy or pleasure, and not to overwhelm myself in the preparation of it, then I could summon up that emotion. Having the right people around me could conjure this. I wanted to express those feelings of love and gratitude that had been absent for so long. I felt as if I had been away, totally absent, for a long time, and was so relieved to find everyone still there when I returned that I had an urgent need to gather them around me, hold them close and thank them for waiting for me. And the only way I know how to show people I love them is through food. So I cooked.

Taking time to make a meal for someone is a gesture of love. Even if that person is a stranger, you have still poured your time, ideas and consideration into a plate of food and presented it to them, like an offering, for their enjoyment. Now, if you have gone to the extra effort of *growing* that food as well, if you have nurtured that parsnip from seed, a seed you have tended for the past eight months, then presenting someone with a plate of parsnips, roasted with miso and sprinkled with seaweed, is a surprisingly loaded gesture.

I can't deny there was an evangelical glint in my eye during those meals. My new church, the natural world, had claimed me as a preacher, and I wanted to show everyone the wonder of nature. My friends thought I was feeding them a stuffed pumpkin, but what I was really saying was, 'Look! Behold! This pumpkin was but a seed the size of my little fingernail a few months ago, and see it now – huge and glorious – a miracle of nature. Worship at the altar of this stuffed pumpkin, for it is life itself.'

I wanted to convert them, wanted them to witness this wonder, to feel the same awe that I had found so restorative. This fanatical enthusiasm would prove useful one day.

Kathy Slack

Stuffed Pumpkin to Convert

A stuffed pumpkin is a cheap trick, really – so huge, so theatrical, so generously welcoming, it is bound to convert even the hardest heart to the miracle of vegetables. Its divine scale in the centre of the dinner table is so wondrous as to virtually guarantee a new disciple before anyone's even eaten a mouthful.

The knack to stuffing a pumpkin is to take a wide, round pumpkin, slice the lid off like a boiled egg, scoop out the seeds and place both pieces, cut sides up, in a roasting tin. Rub the pumpkin, inside and out, with olive oil, season with salt and pepper and roast at 200°C until cooked through, about an hour.

The filling is best begun with a grain. I favour pearled spelt or barley; oat groats are fun (from Hodmedod's UK grain farmers), but freekeh works too – something that won't go mushy once cooked, so no lentils or beans (unless you use carlin peas, also from Hodmedod's, which retain their bite). You'll need about 250g, cooked and drained weight, to fill a medium pumpkin and serve 4–6 people.

These grains can be combined with sweated shallots/onions/leeks/celery/garlic (any or all); something fried and deeply savoury – could be pancetta, sausage meat,

mushrooms (a dab of miso paste helps enrich the latter);
some earthy herbs like sage, rosemary or thyme; and
finally, a few nuts or seeds (walnuts, hazelnuts, anything
autumnal) for crunch. Mix it all together and taste – it
should be so delicious you have to go in for seconds.
If not, add more umami (say, miso) and salt.

Stuff the warm pumpkin with the filling, pop the 'lid'
on, and return to the oven for 20 minutes to warm
through. Carry the whole miracle to the awaiting
congregation at the table and watch them succumb
to its glory as you serve up.

You can also make a slightly less flamboyant version,
and an accidentally vegan one, using a shop-bought
butternut squash. As follows:

SERVES 4–6

1 butternut squash
4 tbsp extra virgin olive oil
500g chestnut mushrooms, finely diced
2 tsp brown rice miso paste
1 onion, finely diced
2 celery sticks, finely diced
2 garlic cloves, crushed
8 sage leaves, finely chopped
1 tbsp chopped thyme leaves
60g chopped toasted hazelnuts

50g pumpkin seeds
250g cooked wholegrain spelt (or any other grain)

Pre-heat the oven to 200°C.

Halve the butternut squash down the middle from top
to bottom. No need to peel it. Remove the seeds, drizzle
with 2 tablespoons of extra virgin olive oil and season
generously. Roast for 45 minutes to an hour, until soft
and caramelized on top. Scoop out some of the soft
flesh to make two boats. Put the flesh to one side
for later.

Turn the oven down to 180°C.

Set a large frying pan over a high heat and fry the
mushrooms with a big pinch of salt for 5 minutes until
much reduced. (The salt will bring the water out of the
mushrooms, so there's no need to add oil, it only makes
the mushrooms greasy.) Stir in the miso.

Turn the heat down to medium-low, add the remaining
2 tablespoons of olive oil together with the onion, celery,
garlic, sage and thyme and fry gently for 8–10 minutes
until the onions are soft. Stir in the hazelnuts, pumpkin
seeds, spelt and the reserved squash flesh and mix well.
Check the seasoning; it'll likely need pepper.

Now spoon the mixture into the hollowed-out squash
halves. At this point you can put the stuffed halves in the
fridge until needed (an absolute winner if you're after a
make-ahead vegan dinner), or return immediately to the
oven for 15–20 minutes to heat through.

Cooking, and eating, connects us with nature. In *Nature Cure*, the writer Richard Mabey talks about writing and culture as an interface between humans and nature:

> I believe that language and imagination, far from alien-ating us from nature, are our most powerful and natural tools for re-engaging with it … Culture isn't the opposite of or contrary of nature. It's the interface between us and the non-human world, our species' semi-permeable mem-brane.

I think the same is true of cooking. Cooking local, seasonal food connects us with the nature around us that created it. It is, literally, the natural landscape on a plate. Nature is not something you go and visit in the shires, it is right there in your fridge, kitchen, shopping basket, or on your plate at sup-pertime. We often forget this when we buy vegetables in the supermarket, washed, trimmed and wrapped in plastic. Their uniform shine seems so far removed from the open fields they grew in (if, indeed, they did).

That December, it was not the season to be in the vegetable patch every day, but I was still engaged with nature because I was cooking my harvests. Growing food and cooking are part of the same process for me, and both connect me to nature.

Once again, the vegetable patch is quietly lifting me out of my melancholy. It is reconnecting me with nature and encouraging me to nurture both myself and others. At the time I would have argued that I was the one doing the nurturing, but now I realize it was the vegetables that were nurturing me and showing me that I wanted to nourish other people too.

So that Christmas, our table is piled high with harvests from the vegetable patch. I have every meal planned out. There are spinach and egg frittatas for breakfast, or homemade hedgerow jam for those who prefer crumpets (a family Christmas tradition). Lunches are beetroot soups, a ham with allotment chutney, apple crumble frozen after the autumn harvest. Quiet days between Christmas and New Year mean raw cabbage slaw studded with pomegranates, or pumpkins wrestled from the mice in the shed and roasted for risotto.

On Christmas Day, family squeeze in around the table, jostling for potatoes (I found them) and helping each other to braised red cabbage, my cabbage. The parsnips are tiny, but taste all the sweeter for being home-grown, as if you can taste the personal triumph. The meal is comforting and satisfying. And I am happy. I have found none of it stressful. I enjoy everything from picking the crops to planning the menu, cooking it and to washing up afterwards. It is creative, joyful, generous work.

And it plants the seed of an idea in my head.

Sproutiflette

When you have grown your own food, you want it to be centre stage. After all, it is miraculous because a few months ago this was nothing more than a glint in your eye and a packet of seeds. The turkey seems a bit secondary when you know someone has spent ten months coaxing a sprout seedling – that notoriously tricky crop we have established I cannot grow – into adulthood. So, I am often riled when sprouts do not receive the attention they deserve. Or worse, they are actively ostracized, brought to the table because tradition dictates rather than because anyone can be bothered to cook them in an appetizing manner. Because, yes, a sprout boiled for 15 minutes and served up as a soggy, slushy mess will be revolting. But shower them with love, put them in the spotlight and they will repay you with a wonderful winter supper in minutes.

Dairy farmers of the Haute-Savoie would be very put out about my replacing Reblochon, the traditional tartiflette cheese, with Camembert here, but then I figure I've already earnt their contempt by replacing the potatoes with sprouts. Plus, Reblochon isn't always easy to find.

SERVES 1 (JUST DON'T TELL YOUR CARDIOLOGIST)

175g sprouts (trimmed weight), trimmed and halved
80g smoked lardons
10g butter
50g shallots, finely sliced
1 fat garlic clove, crushed
1 tsp chopped thyme leaves
50ml dry white wine
50ml double cream
125g ripe Camembert (½ a wheel)
    or Reblochon if you can find it
Crusty bread and a crisp salad, to serve

Boil the sprouts in salted water for 2 minutes, absolutely
no longer, then plunge into iced water to cool. This stops
them overcooking. Drain and leave to dry out a little.

In a frying pan set over a medium-high heat, fry the
lardons until the fat is released and the bacon browns.
Fish out the lardons with a slotted spoon and set aside,
leaving the fat behind.

Turn the heat down a touch, add the butter and,
once it fizzes, tip the shallots into the pan, sweating for
10 minutes until soft but not browned. Add the garlic
and thyme to soften for a couple of minutes.

Now pre-heat the grill to medium.

Add the wine to the frying pan and bring to a bubble,
simmering until almost totally evaporated. Remove from

the heat and stir in the cream, lardons and sprouts and season with salt and pepper.

Tumble everything into an ovenproof dish in which the sprouts can sit snugly huddled together in a single layer. Slice the cheese into 5mm thick pieces and lay on top of the sprouts. Grill for 8–10 minutes until golden and bubbling, then serve with a crisp salad and bread for mopping up. A chilled glass of the wine you cooked with is essential too.

January

it started
with a seed

JANUARY

Imperfection

IN THE VEGETABLE PATCH...

This is no time to garden. It is time to dream. This is time to sit, curled under a blanket in your favourite chair, cup of tea in hand, spaniel at your feet, deciding what to plant and where when spring finally arrives. It is a hopeful, optimistic act, this planning – something I find to be in short supply in January.

And it is important to have a plan. Especially when, like me, you are growing mainly annual vegetables which live for just one growing season and are sown from seed in different parts of the garden each year (unlike perennial veg – rhubarb, artichokes asparagus, for example – which magically return annually). Every year I make a graph-paper map of the beds and mark out what vegetables will go where. I have been known to calligraphy this plan once completed. And sometimes to laminate it. Which might be taking things too far.

Without a plan, laminated or otherwise, enthusiasm takes over and I buy too many and too ridiculous seeds, like an unaccompanied kid in a sweet shop. Faced with a bundle of seed catalogues and no grown-up to hold my hand, my over-excitable mind runs away with itself, pocket money frittered and gone in one giddy moment.

'Sure,' I say to myself, 'I can squeeze in five different types of tomato. Oooo, a heritage-variety melon, that would be fun and probably fine without a greenhouse even though the blurb on the packet says otherwise. And gosh, that organic red quinoa seed looks completely feasible. Wait, okra!'

So distractible am I, so easily hypnotized by shiny new varieties, that without a plan, I would have a vegetable patch full of obscure, temperamental, notoriously time-consuming plants that yielded nothing but a sense of mounting frustration and an empty larder.

I also suffer from space optimism. Someone once told me I had what they described as time optimism. I had never heard of the idea, but it turns out to be the delusion that makes you imagine yourself able to fit more into an allotted time than is feasible. Space optimism is similar. I think I can fit more vegetables into the space I have than is possible. At least not without accessing another dimension.

Having a plan stops me doing these things. The plan will not be adhered to. Not at all. It's largely a work of fiction. I will move things around, or will decide to squeeze something else in when the moment comes. Plants might be bigger or smaller than I remember, and I will have to adjust space accordingly. But adher-

ence to the plan is not the point. The point, partly, is to prevent overbuying, but mostly it is to imagine a possible future that is beautiful and bountiful. To hope that even half of it might come to fruition offers light and sustenance in the January gloom.

So there, by the fire, in the listless, grey new year light, I am surrounded by seed catalogues, attempting to curb the excesses of my fervour with ruler and graph paper.

And it must be catalogues, by the way, delivered by post. Searching online won't do. Browsing seed catalogues is a ritual that growers have enjoyed for generations. Flicking through the Bible-thin pages of a poorly printed, badly arranged pamphlet and marvelling at the order form on the back page, which looks like it has been designed by the same people who make the customs forms at airport immigration, is all part of the charm.

Ugh. New Year. So many resolutions. So many 'shoulds' and 'shouldn'ts'. So many promises and programmes to make a more perfect version of yourself. A commitment to 'live your best life' or realise a 'new you'. So much rule-making. If I could only follow the rules in this listicle in the weekend supplement, then all my problems would be solved. It gets to me at the best of times because I am susceptible to the charm of new beginnings. The promise of betterment is bewitching. There is nothing I like more than a clean slate or the first page of a new notebook. Let's start afresh, it says, let's bin all that old mess we made last year and just *do it better* this time. Just try harder.

But what was wrong with the old me anyway? The old me started out last year trying to be better too. Did she fail? Do I have to do it all again? Is it always a test I am doomed never to pass? Do I still need resolving, improving, bettering, detoxing?

In previous New Years, I would have nodded fervently and asked where I sign up to the diet plan to go with my new gym membership, my Top 100 Books to Read This Year list and the app that records how much water I drink. (It's no wonder this toxic 'betterment' used to make me feel I was constantly falling short.) But this year things are different. This year I am better equipped to bat away the temptations of a New Year. Partly because the vegetable patch has been teaching me to accept imperfection.

But this is also because now I know that this isn't even new year. Not really. Not for nature. And so not for me either. For us, New Year isn't until March, when spring arrives. In fact, many ancient and medieval cultures, wisely, celebrated New Year in late March. It was only with the imposition of the Gregorian calendar that a January New Year was gradually, and often reluctantly, adopted. Spring is when the garden begins its year and life is renewed. This is when I come out of hibernation. In the garden, nothing is new about January – everything is asleep. Me included.

Now, the veg patch feels at its most unkempt; its least perfect. But in truth, things are constantly imperfect in the garden. In

May, the broad beans get blackfly; come July, the pea netting snaps; the mice eat the seedlings any time of year; the weeds keep coming regardless, dishevelling your neatly hoed rows. Once, I lost an entire brassica harvest to a lone cabbage white butterfly that got in under my netting while I was away with work and laid an army of caterpillars, who munched the whole crop. When I returned, nothing but the stumps remained, surrounded by a moat of green caterpillar droppings. Untidiness and imperfection. Everywhere.

Another time, I arrived at the farm one morning, but before I got to the walled garden Benevolent Farmer Brown dashed out of the kitchen and stopped me.

'You'd better come in,' he said.

'Everything OK?' I asked. He looked like someone had died.

'I'm so sorry Kathy, but Blossom and her gentleman guest escaped last night.'

Turns out that Blossom, a pig, was being 'visited' by a male pig in the hope he would sire some piglets. This boar seems to have been quite a charmer because he thought a fancy meal was the ideal way to set the mood for the evening's romance. And where better to lavish his date with fresh, organic vegetables than the garden that adjoined the pigpen. So, Casanova headbutted the pig fence repeatedly until it gave way, escorted a delighted Blossom to the gate of the kitchen garden where he winked and flexed his muscles (one imagines), and then lifted the gate *off its hinges*.

Now, by this point Blossom is already thinking, 'Gawd damn, now *that's* a man right there.'

But he goes a step further, takes her by the trotter (possibly) and leads her into the patch where, after an *amuse-bouche* of sweetcorn, he introduces her to the brightest thing he can see on this moonless night: my huge, orange pumpkins. Swept off her feet, Blossom abandons decorum and tucks in, sampling a single bite of each glorious, sweet pumpkin before moving on to the next. They gorge themselves, mauling the entire crop, before returning to their pigsty, replete and euphoric from their adventure, to... Well, we can imagine the rest because Blossom had piglets later that year. I called them all Pumpkin.

A lot goes wrong in the vegetable patch. But the world doesn't end. Things keep progressing. You still get a harvest, and it is still joyful. It might be a bit knobbly and a bit nibbled, or sometimes a lot nibbled, but it still feels like an achievement.

If rampaging, amorous pigs won't teach you to embrace failure and imperfections, then gardening books will.

Early on in my vegetable growing, I would pore over gardening manuals. Not one to just have a go, I like to be taught how to do it first. I'm a sucker for a class. So, I read every 'how to' book on growing your own food that I could find. And, gosh, they like their rules. It can all seem quite complicated.

For the first few years in my vegetable patch, I would stand over the soil, clutching a mucky, dog-eared manual, reading the instructions (which I had highlighted the night before, naturally), adhering to every word. I had a pH kit to test the soil before planting anything lest it be too acidic or alkaline. According to the books, accidentally planting potatoes in a soil with pH 6.5 when potatoes prefer pH 5.8 would spell disaster. I measured, with an actual ruler, the distance between seeds as I sowed them. If the beetroot was sown too closely packed it wouldn't grow bigger than a walnut – more disaster. I read so extensively about club root, a fungal disease that affects brassicas and stays in the soil for decades making planting anything impossible, that I was too paralyzed by fear to plant any brassicas at all. It wasn't worth the risk, was it?

This is the problem with manuals. They make the whole pursuit of gardening seem fiddly and difficult, as if you need a PhD in botany before you can sow a beetroot. Nothing is more off-putting than punctiliousness, especially to a beginner; it can make you feel like it's all too complicated and cause you to throw your hands up in despair, lock the shed and take up fishing instead.

The big secret that the gardening world will not admit to is this: *no one follows all these rules.* Apart from, that is, the fastidious old duffers on allotments who sit by their sheds passing judgement on your dahlias and worrying about winning the village show with their enormous marrow. These are the same mouldy folk who get exercised about new-fangled inventions like pak choi and Tenderstem broccoli. You must forgive them. They are lonely souls, adrift in a changing world. The rules are their life raft.

Anyway, in the first few weeks that I was off work, when I discovered the veg patch, I was in no state to consult the manual and just chucked a few lettuce and radish seeds in the soil. I didn't measure the distance between each seed. I didn't sieve the compost I covered them with. (Actually, I don't think I covered them *at all*.) I didn't keep a note of when I watered or what sort of soil I'd put them in. I even omitted the painstaking tedium of thinning them. And, do you know what happened? Right first time: nothing. Everything grew *fine*.

Since then, I have adopted a bung-it-in-and-see approach. Which is pretty much how I try to approach life generally these days. Sure, sometimes it doesn't work. I lost all the broad beans one year because I planted them too close together in the hope of squeezing a few more in but instead created the perfect conditions for chocolate spot (a fatal fungal disease). But hey, now we know that spacing matters.

And in many ways, imperfection makes for a *better* world. Rewilding, for example, encourages precisely this: leave a space alone and allow nature to take its course, growing as chaotically, rampantly and indiscriminately as she likes. The result will be imperfect, full of so-called weeds, but it will also be a more stable and resilient patch of land, supporting more diverse wildlife. Now, many argue that, while this is a valid approach to some landscapes, it's no way to run a garden, which is, intrinsically, a space where nature is managed for our benefit. For me, there must be a middle ground, so to speak, in which we can make room for the havoc of nature and benefit from the increased biodiversity and resilience offered by wilder growing methods, but still manage things enough to get a harvest. So let the weeds grow a little. Leave the spent harvests to dry and rot.

Resist the urge to tidy (not difficult). Garden with the intention of it being imperfect. Anyway, gardening with the aim of perfection would surely be a joyless, not to mention doomed, pursuit.

If ever something failed – more mess – I now saw it in a new light: 'Oh well, the world keeps turning. And at least it'll make good compost, or a nice home for bugs.' These were not catastrophes. Just imperfections. And even imperfections have their uses. It is almost never the end of the world.

January is hard for perfectionists. But this year, a summer in the vegetable patch has helped me see that bending with the ebb and flow of life, like a bean cane bending with the gales, will not ruin the bean harvest. As it turns out, imperfection is not a disaster.

By January, the kitchen garden is mucky, bedraggled, mostly empty after the Christmas raid. Limp kale corpses, brought down by those cruel co-conspirators rain and frost, slump over, the weight of their dishevelled heads dragging them down for one last kiss of the frozen soil before death. Some crops have failed, like the frisée lettuce, which I was assured would survive the winter but has, instead, turned to slimy mush. Imperfection is all around.

A Meal from the Frisée that Made It

Should any of your frisée survive the winter, you will
want to honour its fortitude by making it the main
ingredient in a meal. Purists might do this in the French
style by tossing it in fatty, fried pancetta. But I favour
something more complete. I toss the leaves from a whole
head of bitter frisée with a sliced orange, a small, shaved
fennel, 100g fried pancetta (no sense in abandoning all
tradition), a couple of tablespoons of flaked and toasted
almonds with sourdough croutons,* all lovingly drowned
in a mustardy sweet dressing (like the one on page 74).

*For the croutons, tear a thick slice of crustless bread into
bite-sized pieces. Crush a garlic clove into a paste, using the
flat of a knife and a little salt as an abrasive. In a bowl, mix
the garlic paste with 2 tablespoons of olive oil, then add the
bread and toss, making sure each piece is evenly coated.
Arrange the croutons in a single layer on a baking tray
and roast for 7–10 minutes at 200°C, turning over halfway
through, until the bread is crisp and golden. Remove
from the oven and set aside far away from where you are
cooking, or you'll eat them before you finish the salad.

Ordinarily, this level of imperfection would send me into a downward spiral of recrimination. What did I get wrong? Where did I fail? But there is no fixing the mess – too cold to plant anything else and too inclement to get outside and tidy up. And – guess what? The world does not end. The imperfection and failure remain and the world keeps turning. Actually, it looks tragically beautiful. Apt for January. This is what the world is supposed to look like when it's asleep. It's not meant to be pert and perfect all the time. It's all part of the growing cycle. It's nature.

This was a revelation to me. Until then, I had assumed things fell into two camps: perfect or awful. There was no in-between. Yet here was my vegetable patch, imperfect and right at the same time. It makes me reflect on what I've heard more recently called 'the quiet tyranny of self-improvement'. The feeling that things must always be improving; looking more beautiful, more productive, more successful, more, more, more than they were before. These are ridiculous expectations. And in yoking ourselves to this plough of betterment we create a self-made hell of oughts, shoulds and musts which see us forever comparing the past with the present, wondering how we could do better. And this, from within, destroys us.

Why do we inflict these impossible ideals upon ourselves? The veg patch isn't always improving; it just repeats itself year after year. And it is glorious. Nature, and my scruffy veg patch, were exorcising this perfectionism from me. It bothers me even now that it took thirty-five years to work this out and that no amount of life experience could teach me what a dead kale plant could.

By January, I thought I had found a new sense of calmness, of proportion. I was not taking things too seriously and was bending with the ups and downs of life. Hadleigh ran off after a deer on a dog walk one day and I didn't panic and assume he would be lost for ever. Instead, I laughed at his feeble attempts to keep up with a stag. Paul and I had a weekend away, a landmark, and I got food poisoning on a biblical scale. But never mind, there'll be other trips. The boiler broke down and I shrugged my shoulders as I rang the plumber; oh well, it'll fix. All such minor hiccups, but once they would have seen me scurrying, overwhelmed by variables, to hide in a dark room. Or an attic office. I could embrace imperfections. I was more resilient. I felt good.

So I went back to work.

'There's Always Chard' Pie

Whatever failures you have in the veg patch, whatever dies, whatever gets trampled, eaten or attacked, there will always be chard. Always. Chard grows like a weed, unmolested by all but the most sharp-toothed of slugs and oblivious to weather. This makes it, while not the most exciting of harvests (pretty, yes, but I wouldn't describe the earthy flavour as exactly riveting), a reliable back-up in the winter garden. So, when I look at the bare veg beds, unsure what we'll eat for dinner, I often sigh and say, 'Well, there's always chard'.

This sort of quiche uses a lot of chard mixed in a cheesy sauce and baked in filo pastry. Use any cheese you have to hand (it's a great way to use up ends of a cheese board), but Parmesan is a must.

SERVES 6 GENEROUSLY

600g chard
2 tbsp olive oil, plus extra for brushing
3 eggs
2 egg yolks
150ml double cream
100g feta
50g cheddar, grated
60g Parmesan, grated
5 sheets of filo pastry

Pre-heat the oven to 170°C.

Chard stalks take a couple of minutes longer to cook than the leaves. So, cut the leaves from the stalks. No need to de-vein the leaf itself, just remove the main part of the stem. Finely chop the stalks, then roll the leaves up into a cigar shape and finely shred them.

Warm the olive oil in a large frying pan. Sweat the chard stems for 5 minutes in the oil then stir in the leaves, still wet from washing, and cook for 8–10 minutes until well wilted. Transfer to a sieve and leave to cool, then squeeze the cooled chard between your fingers to remove as much water as you can.

In a large bowl, whisk together the eggs, yolks and cream and season generously. Crumble in the feta. Add the cheddar, half the Parmesan and the wilted chard, then stir to combine.

Brush a 23cm loose-bottomed cake tin with some oil. Line the tin with a sheet of filo, leaving any extra overhanging. Brush the filo with oil, then place another sheet on top of the first at a 45-degree angle. Repeat with the remaining sheets until the tin is completely lined and the base sealed. Pour the tart filling into the filo base. Scrunch the overhanging pastry up around the sides of the tart, brush with more oil and sprinkle the remaining Parmesan on top.

Bake for 30–40 minutes, turning halfway, until golden and set. Give it 10 minutes to firm up before you release the pie from the cake tin and serve. It's just as lovely at room temperature too.

February

the spell was
broken

FEBRUARY

Being

IN THE VEGETABLE PATCH...

The mood of a winter harvest is different from summer's. In summer, there is such abundance that you harvest so many courgettes it becomes a chore. I chuck them in everything I cook without a thought; pickle them, stuff them, stew them, soup them, then foist any excess on unsuspecting neighbours. These are generous, copious harvests. In summer the crops seem commonplace. A delight, yes, but a ubiquitous one.

Winter is different. In winter, you look at the cabbages, red, fat and polished in neat rows, and ponder the picking of one for days. You make a plan for braising it, or for shredding it into slaw or krauting it, and only then, only when you have a recipe in mind, do you pick it. You carefully cut it from the stem, leaving a poignant gap in the regimental lines, and carry it home in both hands like a precious ruby. You make it the centre of the meal that evening, lavishing it with attention. You will remember today as the day you ate The Cabbage. You will use any leftover cabbage for a crisp winter slaw in days to come; nothing is wasted. A winter crop is harder to grow and far scarcer than in summer, so we treat it with reverence. The whole experience feels very different. Summer's crops are plentiful; winter's are precious.

Just before I went back to work, I was speaking to a friend about returning to the office. This friend had first-hand and extensive experience of severe depression and she looked at me probingly as we walked Hadleigh that January. She knew this blindness well. Virginia Woolf describes it as 'riding on a flat tyre' but being incapable of seeing or feeling it.

'Is that a good idea?' she asked. 'How do you feel about it?'

Well, obviously, I feel totally terrified. I have work nightmares and wake up at sweating, heart pounding. But that's natural. Isn't it? Anyone would be nervous of going back. Anyone would feel a bit sick at the thought, wouldn't they? Anyway, I'm totally fine. The depression is completely cured and so I've no excuse to sit about any longer, have I? If I can cook Christmas lunch, then I must be well enough to work. Time to buck up and get back in the saddle.

My bosses, who had more sense than me, would only let me back on a part-time basis. I would start on a two-day week and build up to four. I would have just one account to look after rather than the usual two or three. It wouldn't require much travel and it was in my preferred area, FMCG. My new client's product? Cat food.

I hate cats. And cats hate me (but don't cats hate everyone?). I am a dog person. In hindsight, this just underlines how ridiculous the whole situation was. But at the time I was delighted to be given a senior role on a big client account and thrilled that everyone was being so welcoming in spite of my shameful failings the previous year. I would just have to get over the cat thing.

Before my newly re-high-heeled feet could touch the London pavement (and before my feet could readjust to wearing heels again), there I was, back in the thick of it – thinking in PowerPoint, running workshops, shouting into a conference-call pod while shovelling a Pret salad into my mouth.

But, *cats*. I needed to get better at cats. So, I joined cat-lover forums online. I talked to vets about cat behaviour. I lurked in supermarkets to watch people buy cat food. I even went to the zoo to observe wild cats, the better to understand their domestic cousins. I was determined to become the person who knew most about cats. I would be the cat expert whom everyone turned to for cat-related insight; the most useful, most perfect cat guru. I went home at night and hugged Hadleigh closer.

In my first month back, I took part in a workshop about a

new range of flavours. Flavours are a big deal in cat food. Unlike dogs, cats are programmed to search out different tastes. Feed a cat the same flavour every day and it will move out. (There is a reason for this, which is something to do with how they hunted when they were wild, but I can't for the life of me recall it from my muddled memory of this time.) But for the domestic cat, the owner controls variety (excepting supplements from unsuspecting voles, birds, or indulgent neighbours). So, the cats have a functional need for variety, but owners have an emotional need for it just as much because they see their cat losing interest if they don't rotate flavours enough and think they are failing the cat. It can get quite geeky. The point is: flavours matter.

But what you call those flavours matters more. You have to make the owner feel like they are lavishing love on their kitty by showering it with appetizing options and thus ensuring it won't abscond next door for supper. And this is how I ended up in a workshop generating cat-food flavour names, wondering if Tuna Casserole was more enticing than Tuna Bake and thinking that discovering the preference for the word 'stew' over 'medley' among cat-loving old ladies in Hungary was one of my more brilliant insights (it's a goulash thing).

The depression snuck back. Not because I was worked too hard or left unsupported. In fact, the agency handled me with kid gloves – regular check-ins, careful diary management from our team PA, all the budget I needed for continued therapy. It came back because it had never 'gone away' to begin with. (I'm honestly not sure it ever really does.) And because I had a stimulus response to my job. The same stealthy fog crept into my brain, making my work slow and my thinking laborious. Colleagues waited for days while I failed to deliver work. I thought team members unreasonable when they asked me, very politely, if I might possibly be able to make a 9 a.m. meeting. I was outraged that a client could waste two days of mine, or anybody else's time, doing something so needless as generating flavour names. The team was just trying to get the job done, a job that made the difference of millions of pounds to the business and so to people's livelihoods, but I thought the whole thing preposterous and unwarranted.

This time, though, something was different. This time I was not primarily distressed about being inadequate. I was distressed about it being so pointless. This time, I did not turn the criticism inwards so much, telling myself I was worthless and useless. I looked outwards instead, questioning the very point of the job in the first place. This time I was not falling apart; I was raging against.

Nothing had changed about the work. I had changed.

The spell was broken. And the vegetables had done it.

Next to life in the vegetable patch, my life in advertising looked ridiculous to me. Compare a day in a cat-food packaging presentation with an hour in the vegetable patch and you see the artifice (if it wasn't already blindingly obvious, which, sadly, to me it wasn't). In the vegetable patch everything is real and immediate; things grow, the sun rises, life happens. It is a tiny example of all of creation at its most fundamental. With your hands in the soil you are connected, literally and figuratively, to the earth, to nature and to the elemental undulations of our world. It is *real*.

Next to this, spending days and exhausting many clever brains to decide whether to call it Tuna Treat or Tuna Feast seemed pretend, not to mention comically irrelevant; like living in a computer game. Every day, you are squished into a steel box with hundreds of other gamers, flung through a tube to a maze of high-rise glass superstructures, strapped in an ergonomic seat and plugged into a console, where you jump through hoops such as Guess the Name of the Cat Food Flavour, the console offering a serotonin-inducing *bing* (or shaky trolley) every time you score. Addictive, but pointless.

Only this wasn't supposed to be an addictive game. This was supposed to be a life. In the vegetable patch I felt like I was in among life. So, what was I in among in Ad-land? It certainly wasn't life, or not the one I wanted. In the vegetable patch I was witnessing The Grand Scheme of Things, and next to that Ad-land made no sense anymore. It was nothing but a construct, human artifice, and here I was letting myself think it was a matter of life and death. Again.

I should say that plenty of people manage to work and thrive in advertising without thinking it a matter of life and death. There is nothing wrong with the career or the industry per se. It's just a matter of how you perceive it and how that makes you feel. The issue for me was that my view of the job had shifted. The vegetable patch had changed my perspective.

Actual Tuna Bake (For Humans)

You'd imagine that so much talk of cat food would put me right off tuna bake for good, wouldn't you? You'd be wrong. On a drizzly Tuesday evening in February, when the train journey had been delayed and I'd arrived home bedraggled, exhausted and ravenous, a tuna bake would be my first choice of dinner. Fifteen minutes from coat off to on the table and made with nothing but store cupboard ingredients, this is a meal for when you have no headspace and zero capacity; when one up from a bowl of cereal is all you can ask of yourself.

SERVES 2 SHATTERED SOULS

Put 150g pasta on to boil and pre-heat the grill to medium while you drain a 160g tin of tuna and a 195g tin of sweetcorn and empty them both into an ovenproof dish. Spoon in a few sundried tomatoes too, chopping them roughly if you can find such motivation. If you have something along the lines of crème fraîche add a couple of spoonfuls now.

Drain the cooked pasta and stir it in with everything else. Grate some (lots of) cheddar on the top and bung under the grill for 3–5 minutes until melted and bubbling.

I should have noticed the change in my perspective on my first day back. I walked into the glossy, glass lobby and saw, on the wall, the artwork that greeted us every morning. It was a piece by the iconic artist and designer Anthony Burrill and it read, in bright block print, '*work hard and be nice to people*'. Presumably Burrill took that to mean 'do the work that matters and look after each other'. I had previously seen it as inspiring, the call sign of a decent company with grown-up values. But, in this context now, even on that first day back, I read it as 'put in more hours and suck up the crap'.

With every day back in the office, I realized that my core values, which were lessons I had learnt from nature, no longer matched the values that were important in advertising. The advertising system relies on you always wanting more, judging the future against the present: your consumer always wanting better than they currently have; you always wanting promotion, awards, the regard of peers. It cultivated want.

But I didn't care about that anymore. I was preoccupied with another sort of cultivation. I didn't want to progress. Not in this world anyway. I didn't want to judge. I was happy just existing.

In the kitchen garden I saw real life. Ad-land was all fakery, a Matrix of constructed value systems and false judgements.

One weekend in February, I was watching Hadleigh potter aimlessly around the veg patch, just as I had watched the bugs amble around in the soil all those months ago, oblivious to external demands or expectations. He was just bumbling around, pleasing himself. Not thinking about what he 'should' be doing, or what he'd do next. Not wondering what I thought of his wanderings or whether he was getting more skilled at sniffing around the garden. He was just *being*. And I realized how much time I had wasted evaluating people and worrying about how they saw me in return. I fretted about where they were on the career ladder compared to me. Worried what they thought of my Ad-land progress – too slow? Too pushily fast?

I'd made a career out of encouraging consumers to do the same too. I'd devised ways to make people think that if they bought a particular product they would be considered 'better' by the rest of society and that this was a good, desirable thing. External validation – the judgement of strangers – was the only currency that mattered.

Here, amid nature, no one, no thing, cared how you passed the time. Hadleigh was just wandering around. No agenda. No plan. No progress. There was no value judgement. Questions like what is his worth? What does he contribute? What has he *done*? What is he *for*? All irrelevant. The dog in the garden just *is*. The bugs in the patch just *are*. That's it.

I realized that I must meet everyone in the world, including myself, just like the dog and the bugs: as the creatures of nature they are, unburdened by any expectations to achieve or prove themselves. There was no place for ranking or comparison. It's not that these would be unkind (though they would be). It's that they had no bearing at all on real life because they are fictional. Like worrying whether David Copperfield would like you if you met him. Or if M would think you good enough to be the next double-o agent. Pretend. Irrelevant.

The problem was, what I now knew to be irrelevant was the bread and butter of my job. And apart from this dissonance, which I felt with a new clarity, the familiar tropes of depression were returning. Once again, it seemed as if my skin had been removed and every touch, every noise was excruciating. My thinking, which oscillated between stodgy and incoherently frantic, was dominated by anger and a certainty that I was letting everyone down because I could no longer stomach the Kool-Aid. I sought out my attic meeting room to hide from the overwhelming noise, email notifications, meetings, trains, traffic, people, presentations. There was too much input here. And I was slipping backwards at an alarming rate. But this time I knew what I needed to do.

I quit. With no idea what to do next.

A Meal for the Day You Eat The Cabbage

On the day that you finally pick the first red cabbage and parade it home triumphantly like the severed head of Medusa, you will want to make it the centre of your evening meal. Not always an easy task with a cabbage, but very possible. Here's how.

Set the cabbage up to braise very slowly in the oven. I do this by finely shredding a small red cabbage (900g) and roughly chopping two Bramley apples. Mix them together in a casserole dish for which you have a lid. Stir in 85g soft brown sugar, 50g butter (diced), 75ml red wine vinegar and 100ml red wine, plus some salt and pepper. Clamp the lid on and place in a low oven (about 150°C) for 2½ hours or until the cabbage is meltingly soft. Give it a stir and a check every now and again.

At this point I divide the mixture in two, freezing half for a side dish another day, and turning the rest into a main. Because nobody puts the first home-grown red cabbage in the corner.

For a main dish that will comfortably feed four very hungry people, add to the half batch of red cabbage:

100g pulled ham hock
250g cooked puy lentils

2 celery sticks and 2 shallots, both finely diced and gently
    sweated in a tablespoon of butter
3 tbsp finely chopped parsley

Stir and serve warm to adoring crowds.

March

the soil bubbling with
potential

Potential

IN THE VEGETABLE PATCH...

March stands on the threshold of a new growing season, the soil bubbling with potential for life and expectation all around: nature's true New Year.

Every job in the garden this month is an act of hope. With each seed sown, each bed mulched, is the promise of the coming harvest, a future meal. As I sow green beans – too early but I can't resist – I imagine an August lunch of warm beans tossed in mustardy vinaigrette with nothing but a baguette and a glass of wine for cover. I can almost taste the peppery kick of olive oil, hear the squeak of the beans in my mouth, feel the condensation of the chilled glass in my hand. Sowing seeds is transportive.

The kitchen table is commandeered for seed trays, stuffed with damp compost and a forest of labels, and home to newly sown tomato, chilli and bean seeds, all of which require warmth to germinate and cannot be planted outside until after the last frosts in May. They (and I) would prefer a greenhouse (it would take another decade to acquire this), and there is space in Farmer Brown's, but I like to keep watch over my seedlings and having them in the kitchen means I can check on them when I come down to make tea in the morning, my pyjama cuffs now always muddy from dipping in the compost. At mealtimes, we shuffle the trays out of the way to make room for supper, ignoring the musty smell.

If March is mild and mellow, there will be peas, radishes, spinach and lettuce to sow outdoors too. These seeds will germinate in anything above around 7°C. But the resident mice are notoriously well organized and send out nightly sorties to collect freshly germinated pea seeds, so I plant peas in module trays and cover them with a cloche – double protection against looting and frost. Smaller seeds are not such a prized booty, so lettuce, spinach and radish seeds sown directly in the soil are left unmolested.

These first few rows of seeds, carefully sprinkled into the soil with chilly fingers and a Thermos flask nearby, are some of the most joyful moments of the year.

By now it will not come as any surprise that, on the day I quit, I came home and went to the vegetable patch to shovel compost. After all, the patch had offered me so many life lessons these past few months. Lessons that I needed to relearn after

too long on the hamster wheel. Last time, the kitchen garden had been a reassurance, gently reminding me that I was
capable, that life could be simpler, that being flexible, embracing imperfection, compassion and taking care of myself
were values worth living by. So, when I returned to the patch
to 'commune with the vegetables', as Paul was now referring
to it, I was looking again for insight, comfort and solace.

But I got none of it. No pat on the back, no 'there there,
dear'. The vegetable patch had other ideas.

In the patch, you create something out of nothing. One single tomato seed can become two kilos of tomatoes for a soup,
a tart, a salad. After that, just one of those tomatoes you have
harvested has the capacity to become a hundred more tomato
plants thanks to the teaspoonful of seeds it contains. Potential.

And then there's the soil: seemingly inert, but in fact fizzing
with life and full of nutrients just waiting to send your little
seed rocketing into adulthood. In every seed packet you open,
every patch of soil you rake, every watering can you lug, every
wheelbarrow of compost you shovel – there is opportunity.
Hope everywhere. The possibility for life. And food. The allotment was frothing with potential to create something where
there was nothing. This is what I saw when I went to the garden
after I quit. Instead of comfort, it showed me opportunity. A
blank canvas full of potential.

Also a blank canvas was my career. I would never work in
advertising again; I knew that much. But beyond that, I had
no clue what to do with myself.

As I shovelled compost onto one blank canvas and thought

about the other, I felt like the patch was saying to me, '*Well, what are you going to make with all this then? There's the chance of something here.*'

Two clean slates just waiting for me to create something to fill them. Potential everywhere. What would I do with it? I had never had an opportunity like this.

Uncharacteristically, I was not terrified. I was excited. I felt hopeful; so much opportunity to create. The last year had taught me to fear less.

The only nerves I had were because I knew how lucky I was. I had been able to pay for the help I needed to get better; I had incredible family and friends who'd supported me back to health; I could escape into nature because it was right on my doorstep; I could quit a job I knew was bad for me without worrying immediately about the financial implications. Few people have this chance. I must not waste this privilege.

I still wrestle with this even now. Mostly it is a galvanizing mantra: don't fritter away the days, I say. Make something. Many people would give their right arm to have this opportunity. But sometimes it turns manic and tips over into a sort of pathological FOMO where I must chase down every opportunity. Which is more exhausting to live with than to endure yourself.

Anyway, it was March, and recovery from this relapse is quicker than the first time. Partly this is because it isn't as intense as last time, partly because I haven't endured it for so long before taking action, but mainly because I know my medicine now and I take it willingly, in great gulps – rest, quiet and the veg patch. I clear the remnants of winter from the plot, spread my compost over the bare soil, rake it neatly, tidy round. It's hard, but peaceful, labour; I am seeing clean slates everywhere – my mind, my career, my veg patch – and it is energizing.

The veg patch, though, was not totally blank. The cabbages survived the winter and I had planted far too many of them (no two-person household needs a dozen cabbages). The leeks finally surfaced too. I cooked daily, using every scrap of what remained in the patch. I started jotting down the recipes for meals I was creating with my harvests and taking idolatrous photographs of those harvests too (I had been into photography at university and now that I had a good subject matter, the interest returned). I like things to be tangible, so I collected them all on a blog (it was 2012, there was no other way then). I called it 'Gluts and Gluttony' because it was all about the gluts I was getting from the vegetable patch and the ensuing gluttony in the kitchen. (Too late, it would become apparent that most people misread it as Glutes and Glutony and assumed I was a gluten-free fitness influencer with super-firm buttocks. How wrong they were. When, years later, the misspelling made it onto a poster announcing the line-up for a food festival I knew it had to go.)

I became obsessive. I collected recipe books; I researched

new growing methods that would extend the season; I wrote a blog every week, taking one vegetable at a time and suggesting two or three ways of using it. I didn't think about what it might turn into, I was just enjoying the act of making something.

Advertising is often billed as a creative business, but in truth there are only pockets of creativity in an ocean of PowerPoints. Usually the lowest common denominator wins. Consumers see hundreds of adverts a day. How many do you remember? Exactly.

Here in the patch and the kitchen, I felt truly creative. I planned the coming growing year with a head full of recipes and ideas for an abundant and blowsy patch where straight rows of carrots were mixed up with splashes of calendula, and nasturtiums billowed over the paths beside tufts of wild fennel.

If I wasn't gardening, I was cooking my harvests. My cookery skills were fine. Dad is a brilliant, intuitive cook who can create a meal from an empty fridge and never follows a recipe. Mum is a buttoned-down baker, a follower of recipes and legendary quiche maker. I grew up with both ends of the culinary spectrum, cooking and eating interesting food. I always ate whatever was put in front of me. As an adult, cooking became a hobby, and it was the reason I was so desperate for a vegetable patch. I wanted my own ingredients to cook with. But as I started making up recipes, I realized there was much more to learn. And we know how much I like a class. So, I went to chef school. I had no agenda at that stage, it wasn't a career move. I just thought it would be fun. And since fun wasn't something I had seen much of for a long while, I just went with it.

I found a school in Devon where I could do a week-long intensive course and then go home to practise, before returning a few weeks later for more. Between courses, I armed myself with the classic 'how to' books – *Leith's Cookery Bible* and *Ballymaloe Cookery Course* – and went through them chapter by chapter, making the basics to learn the techniques before moving on. It was the RHS gardening manuals all over again.

One week, we lived off nothing but sauces (hollandaise, mayonnaise, béarnaise, crème Anglaise and all the other '-aises') as I ploughed through the sauces chapter. The next we devoured vegetables chopped every which way as I learnt to brunoise, julienne, batonnet, chiffonade and all the rest of it.

At chef school, my classmates were mostly cooks in high-end chalets or chefs on private yachts. They were hardcore. One girl told me how her Russian client returned to his yacht unexpectedly one night with a group of 'friends' and ordered a lobster soufflé as a late-night snack. It wasn't on any menu, and she had no warning, but she did it without fuss or, apparently, the use of a wand. Another chap was the travelling chef for a family (they had a different chef for home) and could spin sugar into cathedrals. The children enjoyed it on their ice cream, he said. I was the country bumpkin of the group.

The school itself was professional and traditional. The student rooms had a whiff of residential home about them; slightly institutionalized, with faded paisley bedding and plastic bathmats. The teachers, all men, had been through the regiments from catering college to Park Lane brigades to hotel restaurants in the provinces (probably ones I held margarine-essence-defining workshops in) to teaching.

The cooking they taught was technical, whizzy and complicated, designed to show off the skills of the chef and the discerning palate of the diner. We learnt to ballotine chicken breasts, debone quail, sous vide salmon, make foie gras foams, dehydrate and powder scallop roe (that last one is quite fun).

During one class we made raspberry soufflé. It was the most fabulous, most pink thing I have ever made, and it used kilos of fresh raspberries, pressed and sieved into a velvety coulis. But it was spring. What on earth were we doing with raspberries in March? But raspberry soufflé was what they taught on this class, to every group, so we would make it, regardless of the season. The food was completely without reference to nature, totally disconnected from the land that had grown it. I scowled, cooked it, and thought about how strange it felt to cook without nature as sous chef. Of course, I still devoured the whole unseasonal lot.

The menu choices were set in their ways, even if those ways were gunning-for-a-Michelin-star ways. Back then, you did not learn creativity at this school, you learnt cookery. But that suited me just fine, because my inspiration was coming from the vegetable patch.

We were never marked on our cooking skills; this was too commercial a school to be the sort of course you passed or failed; they wanted you to come back. But we were given feedback. Especially on plating up, which was the only area that wasn't tightly regulated and was open to our creative interpretation.

I have always felt uneasy when food is served surrounded by piped foam flourishes and squares of jelly and ornate tuiles. It makes me wonder how many hands have been on it and how long the plate has sat there being fiddled with before it gets to the table. So I plated up my dishes in a way that I thought was enticing and elegant, but more relaxed.

For example, we made sea bass ceviche. A beautiful, simple technique every cook should know. Rather than pressing it into a tian – a deep metal ring like a pastry cutter – to create a perfectly round disc of ceviche in the middle of the plate, I piled mine, artfully I thought, off-centre with a few micro herbs scattered on top and to the side. I worried that the micro herbs were too much, but it looked appetizing to me.

My tutors delivered their critique with only thinly veiled vim.

'Gosh Kathy,' they said every day. 'That plate looks very … *rustic*.'

It turns out that the technical way of describing this approach is 'rustic'. Of course, what they really meant was *messy* or *unsymmetrical* or *dishevelled*, only they were too polite to say so, fearing I might take offence. But I was delighted. Rustic meant it had simplicity and charm, that it could have been plucked from the fields and piled onto the plate, that it was close to the landscape it had come from. Which was exactly what I loved about cooking. They hadn't meant it to be, but I took it as a compliment.

Cookery school might have been too cheffy for me, but it was robust training. I learnt the fundamentals, and I had fun. Most importantly, thanks to cookery school, I came home sure of my cooking style: *rustic*.

Rustic Sea Bass Ceviche

This is a dish to make when you find yourself by the sea with access to the best wild, line-caught British fish possible. If this is not where you are, make something else. Ceviche lives and dies on the quality of the fish.

SERVES 4 AS A STARTER

¼ red onion, very, very finely diced
4 cherry tomatoes, de-seeded and chopped into eighths
1 small, perfectly ripe avocado, cubed
½ grapefruit, pith sliced off and cut into segments
    (so you are left with wedges of flesh but no pith
    or membrane)
Juice of 1 lime

1 tbsp finely chopped coriander leaves
½ small red chilli, de-seeded and finely chopped
2 small sea bass or red mullet or bream fillets
  (whatever's the freshest and most local),
  skinned and cut into cubes with a dangerously
  sharp knife

Mix everything together with gentle hands and a pinch of salt.

You can serve it immediately or leave it for a couple of hours, so that the acidity of the citrus 'cooks' the fish. For me, this spoils it, though it might make it more accessible for those new to raw fish. I favour leaving it to sit for 5 minutes before piling (rustically) onto plates or into cups and serving.

Now I was a competent cook, and I had a plan. I would cook for a living. *Rustically*. But not in a restaurant (too hectic, not enough sleep). Or grow vegetables. Or both. And maybe write some recipes. Anyway, something like that. OK, I had the inkling of a plan. But it was exciting.

Let's not be coy. At first my cooking and the blog posts that accompanied it were truly awful. The pictures were bad, the layout was ropey, the recipes vague, typos everywhere. But the idea of finding lots of ways to use one vegetable was central from the beginning, and remains so even now. One of my first posts was about leeks. It featured an adapted version of a recipe I had learnt at chef school – carpaccio of sea bream with confit baby leeks – and a far more 'rustic' dish of leek rarebit. They could not be more different in approach, but both are delicious and both celebrate the wonders of leeks, which is, after all, the whole point.

Sea Bream Carpaccio with Leeks

SERVES 2 AS A STARTER

8 baby leeks
5 tbsp extra virgin olive oil
2 tsp apple cider vinegar
1 tsp wholegrain mustard
1 fillet of sea bream or bass, skinned
    (must be still-flappingly fresh)
1 tsp capers
1 cornichon, sliced

Wash the baby leeks and trim the ends as delicately as you can. Warm the olive oil in a saucepan and add the leeks. Leave to mingle over a low heat, lid on, for around 10 minutes or until the leeks are tender. Remove the leeks and keep the oil for the dressing.

To make the dressing, mix the leeky oil with the vinegar, mustard and some salt. Give it a taste and adjust as you see fit. You want a little zing to it, but not so much as to overpower the sweetness of the fish and leeks.

To plate up, slice the fish into gossamer-thin slices. The key to this is a very long, sharp, thin knife and confident strokes. Starting from the tail end of the fish and slicing at an angle towards the tail will help too. Arrange the fish in one layer on the plate and brush on the dressing. Scatter the leeks on top together with the capers, cornichon and a few baby leaves if you're feeling cheffy.

Leeky Rarebit*

SERVES 2

3 leeks
Knob of butter
60–75ml double cream
2 handfuls of strong cheddar, grated
4 chunky slices of manly bread

Finely chop the leeks and sweat in a knob of butter over a medium heat for 15 minutes until the leeks are soft but not browned.

Add the cream and bring it to the boil. Remove from the heat and add the cheese, stirring until melted. Season with salt and a dash of mustard, if you like.

Spread the mixture generously onto the bread. Grill for around 5 minutes until golden and melted. Eat hot. Feel comforted.

*For the record, this is not how I would make leeky rarebit now, but I wanted to include the original recipe here. Now, I would make a stiff bechamel sauce, then add the cheese and softened leeks. Both methods work.

But I still didn't have a job.

Thankfully, my newly found laissez-faire attitude to failure was helpful here and I was able to risk trying things that I never would have done before for fear of rejection.

I sent off a flurry of emails to gardeners in the Cotswolds who worked for estates or hotels or cookery schools that had kitchen gardens. I explained that I had quit my job in order to grow and cook vegetables for a living (I couldn't be more specific than that because I didn't know the plan myself) and asked if they needed help over the summer. I had no experience, no qualifications, no career plan, and they had never met me. I realize that hardly constitutes radical risk-taking, but for me it was a big deal. Ordinarily I wouldn't dream of approaching a stranger and asking them to pay me to do a job for which I was totally unqualified.

The perfectionist in me would never have allowed such exposure to likely failure. What if they said no? What if they thought I was a flaky, naive, overprivileged princess who thought their jobs were just games? In the past, these worries would have stopped me in my tracks. But now, I didn't care. So they say no – no harm done. So they misjudge me – their loss. Nature had shown me how to get out of my own way; to see that progress isn't always necessary, failure is good, and that things are always changing. Plus, I had seen far worse things happen in the past year, so someone rejecting me seemed trivial. Perspective.

It was a long shot, but failure had ceased to matter. The sun would still rise. The world would keep turning if it all failed. I had been through worse and survived. And I knew that having a plan made no darn difference because life never went in the direction you thought it would anyway.

One of those emails was to the head of the market garden at Daylesford, an organic farm shop on the border with Gloucestershire. It is a huge and beautiful farm with idyllic rolling fields full of beef and dairy cattle, free-ranging chickens and heritage sheep. The kitchen garden was (it's much bigger now) a ten-acre plot in the centre of the estate and chock-full of perfect, organic produce. The garden supplies the shop, the on-site café, cookery school, florists and a ready-meal production kitchen. (They have a spa and clothes shop too, but that's all cashmere, not much call for vegetables.) It is an Eden, paradise for the organically minded.

And the head gardener of this paradise said yes, I could go and work for him. Just for the busy period over spring and summer, but yes.

I have been meditating every day, give or take, since I started the antidepressants last July. I use an app on my phone which guides you through a different daily meditation for fifteen minutes, and I have been trying, falteringly, to cultivate the habit of doing it in the morning before breakfast. On my first day at the organic farm, I change the app setting to twenty minutes, as if an extra five minutes might quell my nerves.

Because, boy, am I nervous. It doesn't matter that I have walked into the first day of countless high-powered jobs in situations far more high-octane than this. Walking into that kitchen garden on day one fills me with terror. Partly because I am still recovering; remember, going to a garden centre for scones and tea had been a big deal a few months ago. But also because this is my first real test. The first step towards making a new career for myself. Would the plan, if you could call it that, work?

After much deliberation about which wellies to wear, whether to take a packed lunch and other nervous minutiae,

I drive to the farm, park up in the staff car park for the first time, and head to the market garden.

I follow the sign through a path in an enclosed oak copse where Tamworth pigs roam, and the garden opens out before me. The bubble of butterflies in my chest bursts when I see it and I laugh out loud at the sheer beauty of it. It is huge, a vast field lined with rows of vegetables that rolls away from me down a gentle hill and towards farmland beyond. On my left, a patchwork blanket of square plots; the blue-green hue of kale abuts the rainbow stalks of the chard. On my right, battalions of leeks, 50,000 as it turns out, in perfect rows as if stood on the brink, ready to do battle with the cavolo nero that faces them, equally regimented. My nerves vanish, it is all too gorgeous to be nervous about. I look at the soil, the crops and the beauty and think, this is my new office.

I find the team, five or six people at full strength, in the polytunnels, picking salad leaves. There are other newbies that day too so I'm less of an oddity. The day has a pleasing cadence to it. We pick until 11 a.m., break for tea, move into the nursery potting sheds to plant up chillies until lunch, then spend the afternoon packing leaves into mixed salad bags for the shops and hosing the mud from the leeks that another gang has harvested earlier.

At lunch, we sit around in the mezzanine of the vast barn that houses all the gardening kit, eating our sandwiches and swapping ideas about what to do with a glut of lovage, which is growing like a weed in one of the plots but appears to be unsellable in the shop. (Later I discover why: it is revolting.)

After the first few days I discover the team has quirky little rules, a sort of unspoken code that binds them all together. For example: always bring a packed lunch and always be prepared to present it to the group, explaining what filling you have in your sandwiches and why. The potting-shed biscuit tin must never be empty and there are bonus points for anyone who can fill it with rejects from the bakery. Only feeble amateurs use a trowel to weed; real gardeners use a cheap, short kitchen knife with a plastic handle, the more battered the better. You may wear gloves when working, but you should know that you will be mocked mercilessly and asked when your mani-cure in the spa is booked for. The atmosphere is welcoming, good-natured and joyful. How could it not be among all these beautiful vegetables.

A week or two later, the head gardener slings a scythe into the back of the mule (a butch, four-wheel-drive golf buggy we used for getting around the farm) one afternoon, hops in and disappears over the fields.

'Where's he off to?' I ask.

'The woods,' comes the ominous reply. I decide not to ask any more questions.

All is revealed a few hours later when he returns with a trailer full of wild garlic, which, it turns out, grows in a woodland on the estate. It is lush and green and stinks to high heaven. I have never seen it before but I love garlic, so I am instantly enchanted.

I come home and, that weekend, go in search of my own wild garlic patch. Now that I know what I am looking for, I find it almost immediately, out running with Hadleigh.

The woodland whose boundaries mark out my daily run is misty with morning dew. Badger trails criss-cross the carpet of bluebells as it stretches away into the depths of the wood – gnarled, ancient, held upright by moss. The spaniel clatters about in the undergrowth bothering a blackbird who was just looking for breakfast.

This woodland is not technically public, but the whole village, it turns out, picks the wild garlic which grows at its edge, furtively hopping over the rubble of a decayed stone wall to snatch a bagful, and so we are all bound to silence by our shared culpability.

Picture a village green at suppertime, wafts of garlic coming from every kitchen and residents surreptitiously tucking into a law-breaking supper. That is our village between late March and May. Around this time, we go for supper at a friend's house opposite ours and she serves pesto.

'Oh, is that made with wild garlic?' I ask, feeling like I'm now part of the foraging club.

A shifty look, as if to say 'you're just as much a trespasser as I am' is enough to silence me. (Plus, it is very good pesto.)

Clandestine wild garlic harvests – that's about as rebellious as we get here in the Cotswolds.

I become addicted to it. Partly it is the taste, but mainly it is the extraordinary satisfaction that comes from picking something wild and free and then making dinner with it. For me it has an almost ritualistic element that pays tribute both to nature and the past. I imagine the hundreds of years that 'our' woodland patch has been picked by villagers past. On a misty spring morning, with an impressionable mind, it would be easy to think you could see their ghosts picking the same garlic for their supper just as I do for mine. (Though perhaps the stone wall was still standing back then and the woodland occupied, so maybe it was a more risky form of trespass.)

Anyway, the point is that foraging offers something constant, and calming, in a changing world. Not only do we repeat the actions of our predecessors, but also that nature offers up the same wild garlic harvest year after year, oblivious to the changes around her (including trespassing law). For me it is this that makes wild garlic picking so magical.

I settled into the rhythms of life at Daylesford quickly: the harvesting, the weeding, the ebb and flow of the biscuit tin. The pace of life here matched mine and I wondered if I might find a way to stay for ever.

Wild Garlic Frittata

The first rule of foraging is this: if you wouldn't bet your life on it being what you think it is, don't eat it, because you may well pay with your life. The second rule is never to take the whole plant. But wild garlic is the perfect starting place for a new forager because it is so obvious: big, broad leaves (you can pick fistfuls without killing the plant) that grow in clumps in the shade and smell, very strongly, of garlic. The flowers are white puffballs like an allium, which is exactly what it is. If you have very bad eyesight, you might mistake Lily of the Valley for wild garlic, which would be a pity because Lily of the Valley is poisonous. But it also does not smell of garlic, so don't forage with a cold.

SERVES 2

½ onion, finely diced
20g butter
125g wild garlic, washed and roughly chopped (the
    pungency of wild garlic varies over the season and
    from patch to patch, so if your crop is too strong
    for your liking, dilute the wild garlic with some
    spinach leaves)

3 whole eggs
3 egg yolks
3 tbsp full fat crème fraîche
100g strong cheddar, cut into small cubes
30g grated Parmesan
15g pine nuts

Pre-heat the oven to 180°C.

Find a small frying pan which you can put in the oven and in it fry the onion in the butter for 5–8 minutes over a medium heat until translucent but not brown. Add the wild garlic and cook for a further 2–3 minutes until wilted. Season with salt and pepper. Remove from the heat and allow to cool for a few minutes.

Meanwhile, whisk together the eggs, egg yolks and crème fraîche in a large bowl, then stir in the cheddar, half the Parmesan, a pinch of salt, and the wild garlic mixture. Stir everything together then pour the whole lot back into the frying pan.

Top with the remaining Parmesan and the pine nuts, then bake in the oven for 20 minutes until golden brown and set in the middle. Slice into wedges and serve.

April

new shoots at their
sweetest

APRIL

Enoughness

It is time. Life has arrived. The pea shoots have germinated; firm, juicy spears breaking through the soil in their module seed trays, ready to plant out in the garden. In the beds, the blue-green leaves of the newly sprouted broad beans rest their exhausted first leaves on the soil, a feast for the pigeons were they not shielded by the thorny briar of twigs and canes I have constructed around them. Life in April is delicate so needs protecting, new shoots are at their sweetest, most succulently tempting, like a fairytale princess.

It is time to sow too. Everything. Since, in the Cotswolds, we aren't clear of frost until late May, I am careful about sowing tender vegetables outside and mostly keep them cluttering up the kitchen table for another month. The courgettes, sweetcorn, calabrese (broccoli) and squash all fight for space among the tomato seedlings here. But outside chard, spring onions, lettuce, peas, radishes, spinach, kale, beetroot – in fact a seed catalogue full of vegetables – all get sown directly into the soil.

I resow the green beans I planted indoors last month. Every year I sow too soon, unable to quell my enthusiasm, and end up with a tangle of leggy, matted vines desperate to go outside, but unable to withstand the spring chill. So I keep them too long indoors on the windowsill and they grow, knotted and gangly, up the window frame, into a mass of growth so fresh and brittle that it snaps at the slightest touch and the seedlings are quite impossible to transplant to the beds. I once tried cutting back the tendrils of these overeager seedlings to allow them to re-grow, but found they just sulked in their module seed trays like a freshly shorn child in a barber's seat, and refused to grow. Every year I resow the beans and promise myself I won't plant so early next year.

I also resow some lettuce seeds. Given the moments it takes to sow lettuces, it's worth giving an early planting a go, as I did in March, but be prepared to resow in April. Many March sowings will not germinate, and those that do are often food for the mice or are scuffed up by foraging blackbirds. But since one packet of seeds contains a thousand seeds, it's no great cost – just the price of sending up a little prayer that the mice have already had their fill.

Most of the manuals will tell you to plant different potatoes at different times: beginning with first earlies (new potatoes) in mid-March, following with second earlies (also new potatoes) in mid-April, and then maincrop (old potatoes) in mid-to-late

April. But we've already established the problem with manuals, so I plant the whole lot anytime in mid-to-late April. That's if I grow them at all. A potato, in my view, is a potato, whether shop-bought or home-grown (controversial, I know) so I often don't bother and save the space for something more interesting. If I do find space for potatoes, it will be for a row of something interesting, like Anyas, which are nutty and creamy. I will admit, digging up their caramel, pink-flecked tubers in August is like digging for diamonds. I do it more for the joy of harvesting than the eating.

A Salad for an Interesting Potato

SERVES 2 GENEROUSLY (OR 4 AS A SIDE)

300g Anya potatoes
150g broad beans, podded weight
2 egg yolks
1 tsp apple cider vinegar
Up to 100ml of your best extra virgin olive oil
6–8 radishes, trimmed and halved
Small handful of lambs lettuce, baby spinach
   or salad leaves
4 spring onions, finely chopped
1 tbsp finely chopped parsley
1 tbsp finely chopped mint

2 tsp baby capers

3–4 cornichons, sliced

Place the whole potatoes in a pan of cold, salted water, bring to the boil and cook for 15 minutes or until a skewer pierces the middle without effort. When the potatoes have just a couple of minutes to go, add the broad beans for 2 minutes. Drain and tip into a serving bowl, halving any large potatoes as you do.

Meanwhile, make a mayonnaise. Put the egg yolks in a bowl and whisk in the apple cider vinegar and a pinch of salt. Ever so gradually, whisk in the extra virgin olive oil, pouring it in a thin stream so the mix doesn't separate. Yes, you can do it in a food processor, but where's the romance in that? Spoon the mayonnaise into the bowl of warm potatoes.

Add the radishes, leaves, spring onions, herbs, capers and cornichons and toss gently. Serve while still warm.

While I was working in the kitchen garden at Daylesford, a permanent job came up in the farm's cookery school. The role was for an all-round assistant – taking bookings, welcoming guests, helping out with the cooking, washing up, doing tours of the farm and anything else that the tutors needed.

The idea of working in a cookery school hadn't occurred to me before, but the head gardener suggested I apply, put in a good word and I got the job. I would spend almost four years at that farm, learning to cook, grow and teach before going free-lance. I didn't know it at the time, but this was the beginning of a new life, a new passion and a new career. And it all came from one little email, dashed off in a fit of optimism, and the perspective that I had found in the vegetable patch.

Someone, ungratefully now forgotten, once told me that a good life could be had by jumping from one lily pad to another like a contented frog. Their point was that the frog doesn't worry about whether the next lily pad is an improvement on the last one; she just thinks about whether she likes the look of it, whether it will hold her weight, fit her needs. And if it will, she jumps.

I liked the look of the cookery school lily pad so I jumped. I didn't stop to think about how it might look on my CV, or what it might lead to afterwards, or if I could see a clear career path mapped out by taking this role. I just liked the idea of it. It sounded fun. It covered some bills, just. It didn't matter if was a progression from my last job. I didn't need to be constantly looking upwards because I wasn't climbing a ladder anymore. I wasn't looking for approval or external validation from any-one. I was on the way to being self-sufficient both in my self-esteem and in my life on a more practical level; I was doing more for myself, outsourcing life a little less. Most importantly, I was having fun growing and cooking vegetables and someone was paying me a living wage to do it. That was enough.

At the risk of getting tangled up in schmaltzy metaphors, I do think there is much to be said for not thinking of life as a ladder. We do not always have to be looking for the next promotion, the next life upgrade. This is the kind of consumerism, the kind of false hunger, that drives us to live in an imagined future rather than the present. And it makes us ill. Or, if not ill, then exhausted and frayed at the very least, unable to look up from the emails on our phones and see the wood, let alone the trees.

I noticed this one day at the cookery school when my old life unexpectedly collided with the new. We hosted private cookery classes at the school, ripe for company awaydays, and a group of media executives had come from London for a 'bonding day'. The plan was for the group to cook a team lunch, have some drinks, have a tour of the farm, that sort of thing. I was liaising with the CEO's PA who relayed to me, in fits and starts as random edicts were barked at her by the boss, what they wanted the day to consist of.

'Can we make sure there's time for informal chats among the group? He wants to have a few quiet words.'

Later, 'And can there be prizes? You know, for the best loaf of bread. And a wooden-spoon prize for the worst cook. That sort of thing.' It had to be a competition. Of course it did.

The day before the trip, 'Is there broadband and somewhere the boss can do a conference call? He can't take a full day out of the office without checking in, people need him.'

I knew what to expect before they arrived. I had spent a decade with people like this. They arrived from the train at the nearby country station, we met them in the minibus and drove them to the farm, their heads bowed anxiously over their phones on the journey. 'Why is there no signal? Where even *are* we?'

Once we settled them into the school, everyone sitting around the central island where we demonstrated the cooking, they chatted and joked among themselves as we introduced the day, jostling for position and establishing status. Who would sit with the boss? Who was far too integral to the running of the business to leave their phone on silent? Who could be the first to ask for wine instead of coffee and be crowned the 'life and soul'?

We were light entertainment, not important or useful enough to be taken seriously. Some messed around like school kids, steadily drinking their way through our wine stocks as they cooked; others huddled in corners, deep in strategic debates about managing clients, wafting a hand at us to do their cooking for them. This was all fine. We were ready for it.

And then I took them on the tour of the kitchen garden. To be fair, she really pulled out all the stops – clear blue sky, rows of showy tulips in the cut-flower beds, fields of swaying salad leaves in the polytunnels – but gradually, the conversations stopped, the phones were put in pockets, the walking pace slowed, they fell silent at the majesty of it all. A middle-aged man in chinos, sheepishly, asked questions about the crops.

'Oh, my wife grows those on the allotment.' No one ribbed him for having an allotment.

They all tried a radish, gobsmacked that they could just

pick it from the ground, dust off the soil and eat it.

'What? Just eat it straight out of the ground? Who knew?' they said. I smiled to myself. I did, but it took me a while too.

They still pissed about, the bravado peaking in the poly-tunnel seeing who could taste the hottest chilli growing there without breaking sweat, helping themselves to a bottle of rosé for the train home as they left. But the brief moment they took notice, the few seconds they looked up, saw nature, real life, and felt *wonder* – that seemed like a triumph.

It felt familiar too. It reminded me of how altering that sense of awe was and how much it had changed my life. I used to be like these guys, always hustling, always head down, always thinking that what really mattered was The Next Thing.

How different it was now. I felt wonder every single day. Each time I noticed a seed had germinated. Or when I put a plate of radishes on the table knowing those plants hadn't existed two months ago. Or when I walked into the market garden in a morning and stood in awe of the sheer quantity of life it contained. And, just for a moment, these guests had experienced that same wonderment.

If you have ever looked at a spectacular view, or gazed at the night sky, or seen a creature give birth, you too will know how nature can elicit awe. For me, it brings a sense of smallness in a vast universe, of calmness in an ever-fluctuating world, of profound connectedness to everything on Earth. It is a feeling that eclipses everything else. All the worries, all the anxiety, all the grief, are overshadowed, made insignificant by the wonder of nature. It is, coincidentally (or perhaps not coincidentally), the same feeling I get when I meditate and it goes well (I know there is not meant to be any 'well' or 'badly' in meditation, but I mean the split second when your mind is quiet, concentrated on the present and you seem to simply float). Awe, wherever you get it, is a drug, a powerful antidepressant.

More competitive minds would tell this story another way: successful woman with two degrees and a glamorous, high-paid job throws it all away in a moment of madness and now earns a pittance picking vegetables. I am the Tiger Mum's equivalent of the Bogeyman, a salutary lesson to keep your children settled in a nice graduate job lest they waste all your hard-earned private school fees. And sometimes I hear the voice of an old-fashioned headmistress in my head, clutching her pearls and saying exactly that:

'You could have done so much. You were doing wonderfully, you were so far up the ladder and then you just let go of it. What a waste.' Actually, that *is* my old headmistress.

Stepping off the corporate ladder does have disadvantages apart from the salary drop. For a start, dinner parties and weddings can be awkward.

'And what do you do?' asks the person next to you as you take your seats at the table. Where to begin?

In the old days, it was so simple. I could reply, 'I'm a global strategy director at an advertising agency.'

They would nod, maybe look a bit impressed. I would bask

in the warm glow of their approval. Then they would ask about *Mad Men* or whether I made the John Lewis Christmas advert. My answer had a currency because they could understand what I did; they saw my job taking up space in their own lives and so they assigned value to it.

Now, my answer is much more convoluted. 'Well, I grow vegetables. And cook. And write about it. And take some photos of vegetables too. And I teach as well, cooking with vegetables. So, a bit of everything I guess.'

A quizzical look. 'Right, so are you vegan?' Nope. 'And who do you work *for*?' Myself. Depending on the audience, there then follows either shared commiserations about the trials and joys of freelancing or, and especially if the person is a little older, a 'gosh aren't you brave?' or a 'well done you' as if I was Shirley Temple and had just climbed Everest.

'What do you do?' It's a minefield. And my career change alerted me to the nonsense of the question in the first place. Why is my job the first thing anyone asks about? Am I defined by how I earn a living? Is my worth as a human being based on whether I am a brand strategist or a button stitcher? And why must I '*do*' anyway? What if I could just *be*? I am more than my To Do list. I hope I am more than, or at least other things besides, my job. And I hope you are too.

I once heard about a woman who, when asked the And What Do You Do question replied, cool as a cucumber, 'Nothing. I'm purely decorative.' I rather admired this. Perhaps not the self-objectification (though if she chose that spot on the mantelpiece, then who am I to moralize?), but certainly the confidence to refuse, in such mocking terms, to bind herself to what Emma Gannon so neatly calls the 'success myth'. She might as well have said, 'I don't need to *do* anything to be worthy of taking a seat at this table. I am not for judging. Success does not define me and I'm comfortable with that, though you clearly are not. Also, I'm hot, I know it, and I'm comfortable with that too. Any other questions?'

It's not an edifying thing to admit to, but I made some big life decisions based on how I would feel in the moment when I had to answer that question. It seems absurd; you can't build a life around managing that one awkward social moment with a stranger, but I did try for far too long.

Having experienced how loaded 'What do you do?' can be, I have attempted to up my game with alternatives. But it's tricky. 'And what do you like to do?' works sometimes but sounds rather like a *Blind Date* question. 'Have you come far?' won't do unless you are royalty. 'Are you local?' works in the countryside sometimes, especially at village gatherings, but can come across a bit 'Come 'ere often, luv?'. I'm coming to the conclusion that there's no good way of starting a conversation that doesn't sound like either an implicit judgement or a pick-up line. But I'm so delighted when, however cack-handedly, people try to avoid the And What Do You Dos that what follows cannot fail to blossom into a pleasant conversation, and I imagine most people feel the same. Unless you're a brain surgeon. Or a UN peacekeeper.

The other issue with stepping off the ladder is that you have to rethink your notion of success. In freelance life, there are no clear goals to reach that will see you progress to the next level. Life will not be linear, climbing ever upwards, getting more recognition with each rung. You will not get a pat on the back for having made it to the next level of the computer game. You will not pass 'Go' and collect two hundred pounds with every turn of the board.

In this vacuum, it is easy to get distracted by false metrics. With no promotion board, no annual review or pay rise, you might look to other measures of your success. Not all of them healthy. Your income, sure, but also, and related to that, how many features you got commissioned, how many Instagram followers you have, how quickly your supper clubs sell out, how well your book sells. All of these are external validations in their own way. Yes, they have real-world financial implications, but they are bestowed on you by others. And they are dangerous because they make success about the volume of regard you have from strangers.

Instead, success could be internally measured. Do I enjoy the work? Do I think what I make is good quality? Does it give me enough time to do the other things in life I enjoy? Do I find it interesting? Success can be measured by connection too: am I doing something worthwhile, even if this is just bringing pleasure to someone's day? Have I added something useful, interesting or inspiring to someone's life?

Once you can pay the bills, the currencies of pleasure,

time, fulfilment and connection can become just as valuable an income if we allow them to.

Unexpectedly, the voice of my disappointed headmistress is sometimes joined by the imagined voices of my feminist sisters, the countless women who fought (and fight) for equality.

'Cooking?!' they cry. 'We battled hard to get you out of the kitchen, not back in it! We burnt our bras so you could have a job like the one you just quit. We'd have killed for the opportunities you scorned. In favour of cooking? The domestic? Geez.'

And that's not all. In some circles, my self-employed status, especially as a woman, especially in a creative field, marks me out as a corporate wife, without a *real* job, just playing at work. Working for yourself – which is notoriously unreliable, underpaid and, usually (heaven forbid), born out of something you actually *enjoy* – means you must have a rich husband for the serious stuff like paying the bills.

'If you were a proper feminist, you'd be able to do this job without the support of a steady, second, household income.' The sisterhood shakes its head and abandons me at the kitchen sink.

Am I letting them down? Perhaps. But I made a choice. And that such a choice was available to me, as it would not have been to my mother, is surely the point. Also, as a wise friend once pointed out to me, 'Honey, the sisterhood is not watching. They've got their own troubles.'

Thinking about all this makes me wonder how much more acute it must be for people who really have opted out. I just changed my job. It's not as if I went to live in a commune, sold all my stuff and joined the Socialist Workers Party. I didn't even get a piercing. Maybe I should have. At least then I would have fitted neatly into a pigeonhole. But I'm not one for piercings.

When you first realize that you no longer have the usual trappings of success, or indeed the badges of having fully opted out (must get that piercing), when you fall between two stools and can no longer answer the dinner-party opener with a succinct, desirable answer, you have to remind yourself that this does not make you a failure. I grapple with this still. But, as ever, I look to the vegetable patch for answers.

Let's take purple sprouting broccoli. PSB, as it's known, is an absolute bugger to grow; really technical – nine months in the soil, needs loads of space, susceptible to every pest going – you get the idea. But it is also utterly delicious and crops at a time, February to April, when almost nothing else does. If you can grow PSB, you can, in my opinion, do anything.

I think about the feeling I get when I harvest a posy of PSB. The trunk of the plant is woody and fat, tied at irregular intervals to a stake in the ground to keep it upright, the lower trussing twine already decomposing having been first deployed six months ago. The tree stands a metre high; its head is a fuzzy mass of tight, purple florets and blue-green leaves, like a stout granny with a purple-rinse perm. This old matriarch was nothing but a seed nine months ago and today, thanks to my nurturing, I can reach into the crown and snip off a bouquet of buds as generous as any from a florist. I carry it home exultant, an armful of potential: roasted purple sprouting with scallops, with feta, with chilli; or steamed and stirred through noodles and soy sauce; or piled on top of cheesy polenta. There was nothing. And now there is a feast. Doesn't that feel wonderful? It does. Aren't you glowing with the glory of your achievement? I am. Isn't it enough? It is.

I compare that feeling to the feeling I got when I achieved something at work – a promotion, a pitch won, a perfectly named cat-food flavour. Then, I felt relief and exhaustion. Both the PSB and the pitch win were achievements, and one is not more worthy than the other, but for me they always felt totally different and their effects are polar opposites.

For a start, I don't need to wait for someone to tell me my broccoli is amazing. I know it. I can see it, unlike in Ad-land where achievement, for me anyway, only existed when it had an audience to bestow value on it. Secondly, I am delighted and joyful about my perfect PSB harvest. Whereas, at work, any success came not with joy but with an overwhelming sense of having been given a reprieve, a stay of execution, for now. Finally, with my harvests I feel like I have created something tangible and uncomplicatedly pleasurable – a simple and totally complete satisfaction. At work, in contrast, any achievement was tinged with a feeling that the success was already in the past, that I was only as good as my last win; a

complicated and fleeting elation tinged with loss, like winning a battle – much was sacrificed, you were left battered
and exhausted, but the outcome was winning, whatever the
price.

So, when the sisterhood and my old headmistress pop by
for lunch (imagine, what horrors!), I can happily tell them
that no, in the traditional sense I am not winning. I am not
the wealthy, high-flying, independent woman I might have
been with a glamorous job and a nice house in Holland Park,
and yes, that might seem like failure to them. But, I can point
to my perfect purple sprouting broccoli harvest, show them
that I am far happier than I would have been if I had become
that other person, and then offer them some broccoli and
anchovy salad for lunch. Like this.

Charred Purple Sprouting Broccoli and Kale with Anchovies and Spelt

SERVES 2

150g wholegrain spelt
150g Russian Hunger Gap kale (or any kale)
3 tbsp extra virgin olive oil
½ tsp chilli flakes
Sprinkling of garlic granules (yes, I know, very seventies)

175g purple sprouting broccoli
¼ red onion
Splash of sherry vinegar
10 tinned anchovy fillets, 4 of them finely chopped
Handful of toasted flaked almonds

Rinse the spelt then cover with cold water and boil for
20–30 minutes until just soft. Drain, rinse again, then
set aside until needed.

Pre-heat the oven to 170°C.

Remove the stalks from the kale and tear it into pieces
roughly an inch square. Pop the leaves in a bowl with
1 tablespoon of extra virgin olive oil, salt and pepper,
the chilli flakes and a few garlic granules. Massage the
oil and seasoning into each piece of kale, making sure
everything is nicely coated but not swimming in oil. Lay
the kale out across two baking trays so the leaves are well
spread out, then bake in the oven for 10 minutes, turning
halfway. Remove when crispy but not brown and set aside
until needed.

Get a griddle pan nice and hot, drizzle a second
tablespoon of extra virgin olive oil over the PSB spears
and then griddle them until just soft and nicely toasted.
You will need to do this in 2–3 batches.

Finely slice the red onion and put it in a large bowl
together with a splash of sherry vinegar, the 4 finely
chopped anchovy fillets, the final tablespoon of extra
virgin olive oil and a pinch of salt. Muddle it all together

and leave for a few minutes so the onion can soften.

To assemble the salad, toss the spelt in the onion mixture, then add the PSB, the remaining whole anchovy fillets and the almonds. Adjust the seasoning if needed and then pile onto a plate, adding handfuls of crispy kale as you go. Serve promptly before the kale loses its crunch.

May

soon to be obscured
by greenery

MAY

Identity

IN THE VEGETABLE PATCH...

If you ever need reminding that nothing – personality, appearance, identity – stays the same for long, look to the kitchen garden in May. As April vanishes in a dank haze of mizzle, early May is hopeful but reticent, greener but with a chill in the air. By the end of the month, however, all is changed. The frost risk has passed, the tender vegetables are transplanted outdoors to brave the milder nights, the RHS Chelsea Flower Show is in full swing. The plot shimmers with livid green shoots, all growing so fast you might hear them squeaking and creaking as they race towards the sky. The cloches are off and the floral dresses are on.

I am at my most idealistic in May. Everything seems fresh, full of promise. In this giddy, be-frocked state, and in an attempt to channel my romantic Country Living notions into a tangible use, I often end up sowing annual flower seeds and planting dahlia tubers. Because nothing is quite as bucolic as flowers in a vegetable patch, and now is the time to plant them. In a few weeks, nasturtiums will creep under the gooseberry bush like intrepid explorers, caring not a button for the thorns; egg yolk orange marigolds will drip onto the paths beside dahlias which stand pert and outrageous, swishing their enormous blooms alluringly as if to say, 'Darling! I'm over here! I know. I'm fabulous, aren't I?' I grow them all, in messy haphazard glee, because it makes me feel like I'm in the Lark Rise to Candleford *stories: a halcyon idyll. I think the feelings they evoke, a nostalgia for a never-known rural past, is as much a reason to grow them as their aesthetic beauty.*

Plus, many can be eaten. Nasturtiums and marigolds especially have a peppery kick that brings colour and life to a green salad. But most are just for the sheer joy. Picking an armful of confetti – cosmos, snapdragons, dahlias, cornflowers, sweet peas, phlox – is rapture to rival the pumpkin harvest.

Everything, flower or vegetable (give or take) can be sown in May. More contained souls than I do this by 'succession sowing'. This is the Buddhist-like practice of planting a little and often, sparingly sowing just enough of what you need each month so that you'll have a constant, but not overwhelming, supply of a particular vegetable, rather than emptying the entire seed packet onto the soil all at once and ending up with a glut that is ready all at the same time. Succession sowing is particularly suited to quick-growing crops that need picking as soon as they are ready – radishes, peas, lettuce, spring onions and so on. I am terrible at it. Truly hopeless. Come May, finally back in the patch after the long

winter, overcome by giddiness I sow far too much and all at once.
I almost never have the restraint or patience to plant a quarter of
a row of radishes every week in the month. I gave up chastising
myself for this failing when I realized that the ensuing gluts were
my main source of inspiration. Faced with a kilo of radishes, I
have found all manner of ways to use them – roasting, pickling,
frying, shaving – which would all have remained undiscovered to
me if I only had a single weekly bunch to unimaginatively dip in
butter or add to salads. Gluts: the mother of culinary invention.

This time last year life was very different. Last May, I was flying
to Beijing, New York, Delhi and Puerto Rico in a vain attempt
to corral a fractious band of clients into agreeing on the brand
essence of a chocolate bar. My days were a tangle of delayed
trains, airport lounges, conference call pods, jet lag pills, Pret
salads, diary clashes, dry-cleaning deliveries and weak hotel
coffee, all illuminated by the persistently blinking red light on
my BlackBerry. My mind was fracturing but I was oblivious to
it, too pre-occupied by the urge to make the wheel turn faster.

My sense of identity and my self-esteem were in the hands
of strangers. Had a client given us their business? Was I progressing at the same rate as (hell, who am I kidding? – faster
than) my peers? Did other people think I was good/smart/
attractive/useful/powerful/likeable? Unless I heard it from
someone else, I didn't believe it was true. I was at the mercy of
external factors.

Today, I am walking Hadleigh through the woods behind the village where the wild garlic is flowering, before I go to work at the farm. I'm wondering what to write about my radish glut and whether I can work wild garlic into the recipe too. The vegetable patch is filling up with plants, the finely tilled compost soon to be obscured by greenery as my toddler courgettes put on an adolescent growth spurt.

People other than my parents are reading my blog. The numbers are low but I don't mind, the making of it is happy work. One local reader has asked me to cook for her summer dinner parties, which seems like a good idea for a business and I start dropping 'private chef for hire' leaflets through some nearby holiday-cottage doors to drum up some more custom. And I have managed to coax a local food magazine to let me write a monthly column about seasonal food for them. Emboldened, I pitch to a couple more local papers to see what happens.

I am working full time at the farm, learning more every day about organic farming, and becoming increasingly convinced that the way we eat is one of the biggest choices we can make for sustainability and for nature. In the cookery school, I am taking most of the tours around the market garden these days. I have hoarded this job for myself and guard it jealously, because I love to see the wonder in people's eyes as the magnificence of the garden is revealed to them; to witness the dawning realization that real life is right here; to see the welcoming tenderness of nature as she puts an arm around them and leads them towards the soil and away from the plugged-in, battery-farmed lives they knew before. I like to see it because it allows me to relive what happened to me and I wish it for everyone.

Life is unrecognizable.

Pit-stop Tart (with Frizzante)

At the halfway point of the market-garden tours at Daylesford, we would pause for a surprise 'picnic'. One of the team would have set off ahead of the group with supplies and prepared a picnic table with a snack and a glass of frizzante for everyone. We'd choose somewhere particularly bucolic and hidden away for the picnic – under a tree, behind the dahlia bed, in the potting shed – so when guests rounded the corner they'd be presented with this unexpected and beautiful sight. It was a cheap trick really, no effort at all, but it always made everyone smile.

One of our favoured snacks for this picnic was courgette and feta tart, because it looks spectacular but can be made while doing five other things (which was always the way at the cookery school). It also goes brilliantly with fizz.

Take a 320g sheet of ready-rolled puff pastry. Score a

1cm wide border around the edge with a knife. Brush the
middle with a roughly 1:1 mixture of grated Parmesan
and crème fraîche (80g and 4 tablespoons respectively
should do it), then arrange three courgettes, sliced thinly
into rounds, in a single layer with some overlapping
(like fish scales) over the top, taking care not to cross the
border line. Drizzle with olive oil and season with salt
and pepper. Bake at 200°C for 20 minutes, then top with
crumbled feta and basil leaves, slice and serve.

I look back on those jet-lagged days in Ad-land and realize I
had been trying to squeeze myself into the shape of what my
perfectionist ideals considered to be a 'successful woman'. It
was a cookie-cutter identity that I thought I had an obligation
to adopt given the privileges I had been afforded. Having a
high-flying career, being exceptional at everything I under-
took, was the only grateful thing to do. It was a duty, a return
on investment, paying back for such good fortune.

Only when I got out into nature did I realize that these
cut-outs were completely fictional – a blow-up doll inflated
by all the hot air of society's expectations. Mostly impossible,
often contradictory, dangerously easy to internalize, but all a
construct. A false ideal.

When I sat on the edge of the raised bed and put my hands
in the soil all those months ago, the constructs of society began
to fall away and, here in the real world – by which I mean in

the natural world – I realized I was fine just as I was. I felt as if the garden was calling me into her world, welcoming me just as she found me, and inviting me to join all the other creatures who were growing, living – *being*. There was no judgement, no expectation, just the offer to exist, together.

Here too I saw that making myself miserable by striving to fit a false identity wasn't the way of making the most of what I had been given. As I kept telling myself, 'Most people would kill to have what you have.' So why waste it in self-recrimination? Being happy with, and self-sufficient in, my identity *was* making the most of it.

But how do we know what our identity is? I think about this a lot, still. And the answer requires greater minds than mine, and another book at least. Plenty of people, understandably, manage to avoid ever needing to face this question by having jobs that define them, or children. Environment, sometimes self-imposed, can make it impossible to ever have the opportunity, or the inclination, to strip away all the trimmings of your life to reveal what it is that you really are underneath the suit, the house, the accolade, the fame, the happy, healthy child. Do I really like, say, skiing, or do I go because friends do, it sounds affluent and they have kids' clubs?

For me, it comes down to values. By this I mean, partly, what you like and dislike, but more fundamentally I mean what you think is important in life. What ideas, standards, beliefs make you do the things you do and cause you to choose certain paths in life.

I had never thought about my values until I became ill.

Being in the garden amid nature, and having shed the skin of my previous identity, I found that precious opportunity to look at my values without interference, without the white noise of daily life impressing on me what society had decided I ought to want.

Perhaps it's circumstantial, but when I thought about my values, I realized that they were all born out of nature. My sense that we are all equal in the eyes of nature; that deep connection with a few other creatures is more valuable than many superficial connections; that our actions, however small, have implications; that kindness and consideration for each other is all – every word of it comes from nature. Say them all together out loud and, yes, they might seem trite at worst, deeply obvious at best. But that they were a revelation to me just illustrates how lost I had been. It seems unbelievable to me, even now, that sowing carrots taught me this about myself.

On a less prosaic level, your true identity, I believe, is also defined by your potential, your strengths. And the veg patch also showed me what I was good at and what I wasn't. Creativity, sharing, making stuff – yes. Neatness, patience, carpentry – not so much.

Only once I worked out what my values and strengths were could I build a life that suited me.

It's important to say that I have not developed these values since I left my old life. They were always there. It's just that the veg patch, nature, made me see them more clearly and realize that the reason my previous life had made me so unhappy was because it didn't match any of my values. In fact, it was totally contradictory to my values and strengths. Over time, living with this discordance became so exhausting that my body and mind gave up. Who can blame it?

Now, when I consider who I am, what my values are and what am I *for*, I know: my identity is a nature lover, a grower, a cook, an eater. Always will be. Regardless of whether I am a high-flying CEO or an unemployed recluse. This is who I am. My sense of self-worth is no longer entirely defined by others, or the status my job gives me, or the trappings of success. It is defined by my connection with nature and the values I derive from that – kindness, compassion, empathy, and an always-full trolley of vegetables.

Grain and Greens Salad

Around this time, I started developing culinary earworms. I have them still. I think they're a good sign – the mark of a cook overexcited about each new discovery. A culinary earworm is when you realize you are repeating an ingredient or technique in everything you cook. You can't get the idea out of your head and keep reinventing it in different ways, often without knowing it. Vegetable hummus – pea, beetroot, pumpkin…; roasting beans (butter, cannellini, anything really) in olive oil until blistered and crispy; topping everything with every possible flavour of pangrattato; suet dumplings of every type on every stew imaginable. I've been through them all.

One of my first earworms was grain-based salads. They were terrific vehicles for whatever was glutting in the garden at the time, and I was excited to discover how many interesting and delicious grain-type-things there were to cook with – barley, spelt, naked oats, but also freekeh, quinoa, bulgur wheat.

It's a very adaptable meal, more an idea than a recipe and simply involves mixing cooked grains (whatever you like) with greens like steamed broccoli, avocado, peas, baby spinach leaves, spring onions, etc. The key is to consider the variety of textures and flavours – creamy, mild avocado, the sweet pop of peas, the allium hit of spring onions. For more crunch, add seeds (pumpkin here) and maybe some toasted almonds. Dress with either simple lemon juice and extra virgin olive oil, or take a more umami-led approach, as described below.

Here's an example for those who like more guidance:

SERVES 2, GENEROUSLY

250g pouch of cooked wholegrain spelt
125g broccoli florets, cut small and steamed for 3 minutes
100g frozen peas, defrosted
1 avocado, sliced

2 spring onions, finely chopped
1 tbsp pumpkin seeds, toasted
1 tbsp sesame seeds, toasted
1 tbsp roasted peanuts, chopped

FOR THE DRESSING:
2 tbsp sesame oil
2 tsp tamari
2 tsp mirin
1 tsp Chinese rice vinegar

Combine all the salad ingredients in a bowl. Whisk together the dressing ingredients, then pour over, toss and serve.

Things I learnt from the soil

It is a decade or so since that May morning walking Hadleigh and pondering my first radish glut. I would work at the cookery school and at the farm for three more years then go freelance, teaching cookery classes and cooking privately for clients who had more friends than inclination to cook for them, or holidaymakers visiting the Cotswolds. As a means to use up my gluts, I started hosting supper clubs, where the menu was determined at the last minute by whatever I had in the vegetable patch. These continue erratically and peripatetically today, but I love the sense of connection they give me. I stuck at the blog until they went out of fashion, replacing it with a weekly recipe newsletter – which are all the rage, for now – and writing for magazines too. I've written a cookbook as well. Each chapter focuses on a particular vegetable. Of course it does, what else would it do? I fill in the gaps with a precariously freelance combination of developing recipes for brands, photographing the food, podcasting, food festivals – whatever allows me to grow, cook, eat or write about vegetables.

That summer when I was first ill, the vegetable patch provided a remedy to the discord in my head. Initially, when my mind was most uprooted, nature grounded me, offering comfort and peace. Then, as I settled into the rhythms of the garden, tuned into the pace of nature, gave myself up to watching the soil, and luxuriated in the simple joy of seeing a seed germinate and then eating the harvest – then I started to learn lessons from nature and rediscover her values. Values of kindness and nurture, of living a more considered, more connected, less judgemental life. Of being enough and having my own agency; of knowing what true success meant and of embracing imperfection. Of creativity and, most of all, of hope.

I have always slightly envied those with a religion – a clear set of moral values to live by that give you conviction in your decisions – but I could never square myself with the supernatural element. What I found in the veg patch, when I put my hands in the soil, when I looked nature full in the face and saw the inner working of the Earth, what I found there was my guiding star, my own kind of faith. Faith that allowed me to know myself better and like what I saw, to be self-reliant, more resilient to the temptations of modern life that so seductively invite us to compare ourselves to others, to judge, to strive for more and to always feel we must improve ourselves to meet a fictional ideal. Nature has become my god. The garden, my church.

Does it matter that my 'cure' was growing food specifically? Could it have been another sort of gardening? Or hiking? Or something else entirely, like knitting? Maybe I would have seen the same issues played out in the trials of darning a sock. Might the tribulations of casting off have taught me the same life lessons? What if I now made a living knitting bespoke jumpers instead of cooking home-grown vegetables?

Perhaps what matters is that I found something that was creative and gave me comfort. Maybe the lesson is to find something that uses your mind, your hands and your heart and make that your life raft, or your lily pad. (I, for example, find knitting terribly stressful.)

But no. That doesn't cover it. Growing your own food offers something totally unique. Something beyond the creativity and comfort of other therapeutic pursuits. It reconnects you with the real world, with nature. And what it has that other outdoor pursuits do not is the power to bestow agency. It gives you the ability to feed yourself and so makes you more self-reliant. Moreover, besides the personal lessons about resilience and perspective, when you put your hands in the soil, when you see a seed turn into a pumpkin, the sheer wonder of it is transformative. Your mind is rewilded, your thoughts freed, you are made anew, your eyes are opened, you are unplugged from the Matrix. And in so doing you are reminded of one very, very important thing: *the sun still rises*. The seeds grow. The seasons change. The rhubarb dies back but is not dead.

Things move on. When you think the sky is about to fall in on you, this will save your life.

Let's not get too dewy-eyed about it, though. There are still ups and downs, though nothing so acute as the first episode. And now I know the signs, I am careful (mostly) about catching myself before things deteriorate.

Because it's not easy, is it? It's all very well finding serenity, self-awareness and enlightenment in the kitchen garden, or surrounded by nature, but it's another thing maintaining that conviction amid the onslaught of daily life. As one fellow student on a nature and meditation course I once took said, 'How do I possibly maintain this state, short of becoming a monk and living in a wood? You've got to *engage* in the world, haven't you?'

Quite. Surround me with daily life, with adverts telling me I'd be cleverer, sexier, fitter, a better woman, better eco-warrior, all round better person if I bought this product; show me an Instagram feed full of #blessed food writers getting their next book deal (in fact, show me *any* Instagram feed); forward me an article about a @FitatFortyFive mum doing her fifth triathlon; a list of the fifty-six Booker finalists' books someone read last year; a diet app that will absolutely definitely work; an ad for a steady job that I really ought to apply for – surround me with any of these, as we all are every day, and I'm still as liable to wobble.

Part of the reason I have written all this down here is to make everything feel more solid. The lessons, the values, the ideas I learnt from the soil are fragile, ephemeral; they evaporate into half memories like dreams do, too easily blown away by the forceful gusts of daily life. Having them in print makes them less easy to forget.

I have some other tactics to help me avoid getting caught up in the weeds too. Spending time growing and cooking food, obviously. But I also meditate, erratically, and when I do I do so outdoors. I go on my silent nature retreat every year. I wild swim (which seems to be a prerequisite of all middle-aged women these days). I bookend my days with micro doses of nature – a dog walk first thing, a ferret around the veg patch at the end of the day to collect things for supper. I avoid London, Instagram, competitions, Amazon rankings, weekend newspaper supplements, a full diary, New Year as necessary because I know they are bad for me. These strategies are all ways to stay close to the wild places that reconnect me with nature and remind me that I am OK just as I am.

But the vegetable patch is still my first defence against any creeping sense that it's all starting to feel like wading through treacle again. I beetle around the patch for an hour weeding, tidying, picking, tying in, digging up and things seem different, shifted, at the end of it.

It would be tempting to try and finish this book by saying that vegetables saved my life. But that would be glib; a neat bow to tie up the story, but an oversimplification. It would negate the drugs, the therapy, the spaniel, the relentless care and patience of my family and friends, all of which made a profound difference. And it might imply that a life-changing and debilitating illness can be cured simply by getting out in the fresh air and growing some lettuce. Or, indeed, that it can be 'cured' at all.

But the veg patch was significant because it was a gateway to something far bigger. It was my way of finding nature, of connecting to it and, in so doing, of finding a new way to see the world and live in it. The simple act of seeing a seed turn into supper was restorative in ways I am only now coming to appreciate.

And I think this would have been the case whether I was growing in a large vegetable patch or in a few balcony containers. You do not need a benevolent landowner with a spare plot to benefit from the joys of growing your own food. Just a few herbs, sown in empty margarine tubs (no, the irony is not lost on me either) on a windowsill can ground you back to nature.

Because, in this little microcosm, whether a patch or a margarine pot, you can witness the magic of a seed germinating and turning into supper. In that single pot, jostling for space on the windowsill, surrounded by old receipts, forgotten post and dusty bills, there is *life*. Life which you have nurtured, weathering ups and downs – aphids, the neglect while you were on holiday, the cat knocking it over (well, if you will have a cat...) – until it grows into a voluptuous basil plant. A green and fervent shrub with an aroma that fills the kitchen, and from which you pluck a few leaves to scatter over your Friday night pizza.

This single plant, that single moment, is a reminder that nature ploughs on regardless. The sun still rises. If things get so dark that you can see no dawn, that little basil plant will be a talisman that encapsulates all the values you hold dear, a reminder of what really matters in life and what does not. A reminder to hope.

Next time you finish a margarine tub, scraping the last morsel onto burnt toast and stuffing it into your mouth as you dash to the office one morning, think about saving that pot and planting something in it. I will send you some basil seeds.

A Very Detailed Guide to Growing Basil

YOU WILL NEED:

6–8 old plastic pots that are wider than they are deep (a wide yoghurt pot, the sort fancy Greek yoghurt comes in, or a rectangular margarine or ice-cream tub are ideal)

A shallow tray for the tubs to sit in

A packet of basil seeds (there are many varieties, but I'd
    start with the classic Genovese, which is the sort you
    buy in the shops), organic and British-grown if you can
Peat-free multipurpose compost (you can use seed
    compost if you're being all RHS about it, but I've not
    found it strictly necessary)
A handful of fine horticultural potting grit or vermiculate
    (optional and, I warn you, the latter, which is an
    absorbent, naturally occurring mineral treated at
    high temperatures and used to protect seeds and
    improve germination, is also the price of gold)
A small transparent plastic bag (a freezer bag,
    for example)
A warm windowsill
Patience

Begin any time between March and June. Use a pencil
or screwdriver to poke 8–10 holes in the bottom of each
of your yoghurt/margarine pots. This is so the water
can drain and prevent the seeds from drowning or
languishing in claggy soil (they are Mediterranean and
not accustomed to damp).

Fill one of the tubs with compost but not quite to
the top, leaving about 3cm unfilled. Tap the tub on
a solid surface to settle the soil and level off the top.
Use a watering can with a rose (the head with little holes
that turns the water from a single stream to a shower)

attached to water the soil until the water comes through the holes in the bottom of the tub.

Open the seed packet and sprinkle 10–12 seeds evenly over the soil. Use a flat hand, or a jam jar, or another tub, to press the seeds down onto the soil, making sure they have good contact with the compost but that you don't inadvertently let them stick to your flattening implement. If you have it, sprinkle over a layer of grit or vermiculite, as thinly as you can, then cover the top of the tub with the open end of the plastic bag, like a hat, leaving plenty of height and air around the soil. Sit the tub on a tray and place it on a warm, bright (but not in direct sun) windowsill.

The seeds will take about 4 weeks to germinate, so sit tight. While you wait patiently, make sure the surface of the soil remains damp. Add a little water to the tray the tub sits in if needed and the soil will draw up. As soon as the basil germinates, remove the plastic bag and marvel at the wonder of creation.

Continue to water (via the tray) until the seedlings are 6–8cm tall and have leaves that look like basil (the first two leaves they will produce are seed leaves and look nothing like the adult leaves, so don't worry that you've planted the wrong thing).

At this point you need to 'pot on'. This means lifting each seedling out of the nursery tub and transplanting it into its own pot. So, fill the remaining tubs (with holes in the bottom) with multipurpose compost and water in the same way as you did before. Make divots in the compost in the middle of each pot around 3–4cm deep. Holding one seedling by its seed leaves, carefully dig it up keeping as much soil and root attached as possible. I use a teaspoon and lift the whole lot up in one heavily laden scoop. Place the seedling carefully into a divot, making sure it is at the same depth as it was in the nursery tub, and gently firm the soil in around the roots. Repeat with the remaining seedlings and return everything to the tray and the windowsill.

Water via the tray as before, weekly should do it, and watch with a full heart as the seedlings grow. When the growing tips are big enough to eat, pinch them out to encourage the plants to bush up. Keep picking regularly and you'll have basil all summer, perhaps even into October. Your pizzas will never taste the same again.

All the Garden Gathered

This dish, or collection of dishes, is called 'All the Garden Gathered' in our house, because it uses pretty much everything available in the summer patch. It is perfect for eating al fresco on the lawn, with a chilled glass of rosé and friends around you. After all, that is how the best meals are eaten: rustically.

Kathy Slack

SERVES 4 (ASSUMING EVERYONE HAS A BIT OF
EVERYTHING)

FOR THE COURGETTE AND AUBERGINE
SALTIMBOCCA:
2 aubergines
2 courgettes
Extra virgin olive oil, for frying
12 slices of prosciutto
Small bunch of basil, leaves picked

FOR THE TOMATO BULGUR WHEAT SALAD:
2 tbsp olive oil
2 spring onions or 1 salad onion
1 garlic clove
1 tbsp tomato purée
¼ tsp chilli flakes
150g bulgur wheat
300g tomatoes
½ lemon, zest and juice
2 tbsp pumpkin seeds
2 tbsp sunflower seeds
2 tbsp pine nuts, toasted
1 heaped tbsp baby capers
2 tbsp sultanas
6 green Gordal olives, pitted and torn

ROUGH PATCH

FOR THE BEAN SALAD:

300g French beans, topped and tailed

1 tsp apple cider vinegar

½ orange, zest and juice

2 tbsp extra virgin olive oil

2 tbsp flaked almonds, toasted

FOR THE SAUCE:

300ml strained Greek yoghurt

75g leafy green herbs (parsley, fennel, mint, hyssop,
    chervil, sorrel…), finely chopped

½ cucumber, grated, and all the water squeezed out

1 tsp apple cider vinegar

Edible flowers, if you have some

FOR THE SALTIMBOCCA:

Slice the aubergines from top to bottom into cross-
sections around 1cm thick. Do the same with the
courgettes but slice them thinner, around 5mm.
You should get around three 'steaks' and two stubby
ends (which can be saved for another recipe) from
each fruit.

Drizzle the steaks with one or two tablespoons of oil,
season with salt and pepper, and fry in a hot frying pan
for 3–4 minutes on each side until browned. Lift the slices
out of the pan and onto a plate lined with kitchen paper.

Lay a piece of prosciutto on a board and place one
aubergine or courgette 'steak' on top at a 45-degree angle.

Arrange four basil leaves evenly on top of the aubergine/ courgette, then wrap the prosciutto over the steak to enclose everything and cover the basil. Repeat with the remaining slices of prosciutto, aubergine and courgette.

In the same frying pan, now set to a medium-high heat, warm the remaining oil, then fry the wrapped 'steaks' for two minutes on each side so the prosciutto turns golden and crisp.

FOR THE BULGUR WHEAT:
Heat the olive oil in a large saucepan over a low-medium heat. Chop the spring onions and cook gently for 3 minutes. Crush the garlic (with a little salt and the flat of a knife) then add that to the pan too, continuing to cook gently for 2 more minutes, until soft but not browned. Add the tomato purée and the chilli flakes and cook for 2 minutes more. Pour in the bulgur wheat, together with 150ml of hot water. Bring everything to the boil then turn the heat off, cover the pan with a lid, and leave for 25 minutes.

Meanwhile, chop the tomatoes into quarters or rough slices. Once the bulgur wheat is done add the lemon juice and stir it with a fork to loosen the grains. Add the tomatoes and the rest of the ingredients and combine. Check the seasoning – it may need more salt, then transfer to a bowl ready to serve.

FOR THE BEANS:

Boil the beans in salted water for 4 minutes, then drain
and refresh in cold water. Pat dry and transfer to a
serving bowl.

For the dressing, put the vinegar, orange zest and juice,
olive oil and a pinch of salt into a jam jar and give it a
good shake. Pour over the beans, add the almonds and
mix.

FOR THE SAUCE:

Simply mix everything together with a big pinch of salt.

To serve, load the aubergine and courgette steaks onto
a big board and take them to the table with the bean salad
and bulgur wheat. Serve with dollops of herby yoghurt
sauce. Make a toast to the harvest and to happiness and
dig in.

Useful things

Gardeners, and cooks for that matter, like nothing better than swapping tips about their favourite suppliers, gadgets and sources of inspiration. So here are just a few of mine.

GARDEN SUPPLIES

DAFEFOOT COMPOST *(Dalefootcomposts.co.uk)*
Rocket fuel for plants, and ethically made from bracken and Herdwick sheep wool in Cumbria. Pricey, but worth it, so save this for your most precious plants.

SYLVAGROW *(Melcourt.co.uk)*
More delicious peat-free, organic compost.

TAMAR ORGANICS *(Tamarorganics.co.uk)*
My first stop when seed shopping. Cornish-grown organic seeds of every sort.

VITAL SEEDS *(vitalseeds.co.uk)*
Organic, British-grown, open-pollinated seeds in beautiful packaging.

CHILTERN SEEDS *(chilternseeds.co.uk)*
A family business offering an extensive range of vegetable seeds, including some oddballs.

JEKKA'S *(jekkas.com)*
Every sort of herb you could ever imagine, and some you couldn't, grown with love and skill by the patron saint of herbs, Jekka McVicar.

HARROD HORTICULTURAL *(Harrodhorticultural.com)*
An online sweet shop of every piece of hardware a gardener might need – from seed labels to fruit cages.

NIWAKI *(Niwaki.com)*
Exquisite garden and kitchen tools. A hori hori (a sharp, Japanese trowel) and a pair of gloves from Niwaki are all a gardener really needs.

FOOD SUPPLIES

PIPERS FARM *(Pipersfarm.co.uk)*
A comprehensive online shop providing a vital route to market for small-scale farmers in the UK. Free-range meat but also a dangerously tempting range of dairy, seafood and store cupboard supplies.

RIVERFORD *(Riverford.co.uk)*
If you can't grow your own food, but want to feel closer to the land, a veg box from Riverford is the answer since they grow so much of what they sell themselves. You can make up your own order if you don't want the pot-luck element of a veg box. And the little newsletter that accompanies each order always offers a fascinating insight into farming life.

DAYLESFORD ORGANIC *(Daylesford.com)*
The mothership. But since they gave me my first job in food, I'm biased. Producing, as they do, their own meat, dairy, vegetables, preserves, cut flowers, charcuterie and more, Daylesford offers the most comprehensive organic food selection. The integrity and ingenuity of the organic farming here has been leading the way for over twenty years.

NAROORA *(Natoora.com)*
Supplier of exquisite fruit and veg (amongst other things) that aims to give small producers access to big markets and help repair the broken food system. It's expensive, but always, always, delicious.

Acknowledgements

I get weepy when I write acknowledgements. It makes me realise how much good fortune is required to bring a book into being: spirited agents, a publisher willing to take a punt, wise editors to guide you, patient and encouraging friends and family to support you. It's a miracle so many books get published really, given that each relies on a set of conditions so unlikely it seems impossible on paper.

My first bit of luck is having everyone at A.M Heath – my agents – in my corner. Zoe King, who put the idea for this book in my head in 2018 and then coached me through early drafts; Rebecca Ritchie – never has such tenacity to secure a deal appeared so graceful; and Tom Killingbeck, whose moment of inspiration landed us with the title.

Their efforts meant I got to work with Robinson and the wider Little, Brown team – as smart, collaborative and lovely a bunch as any author could wish for. Thank you to Emma

Smith who commissioned this book – for the opportunity, the belief and for setting me off on the right path. To Tamsin English, who so deftly took Emma's place as editor: I really landed on my feet there. You are wise and fun and it is a joy to work with you.

And then there are the wider teams – where the real graft gets done: Sarah Thomas, Amanda Keats, Charlotte Stroomer, Clare Sivell (thank you for so patiently indulging me as I pontificated about fonts), John Fairweather, Henry Lord (publicist or magician? I can't decide), Louise Harvey, Aimee Kitson, Charlotte Ridings, Clare Sayer and every last one of the sales and rights teams – proper grown-ups all and generous with it.

I was over the moon when Rosie Ramsden agreed to illustrate *Rough Patch* – she was the food stylist on my first cookbook, *From the Veg Patch*, and is also a talented artist. Rosie, everything always gets more beautiful when you are involved – thank you.

When it comes to thanking friends and family, I've attempted to write all sorts of expressions of my gratitude here and then deleted them because I can't find a way to adequately express my love, my thanks (and my regret for having put you through the wringer) without it all sounding trite or so full of superlatives it becomes empty. To Mum, Dad and Andrea, to Cotswolds and London friends, to old Ad-land pals (notably Howard and Daniel), to wise friends and cheerleaders (the extraordinary Dysons), and above all, most especially of all, to Paul – thank you doesn't do it justice.

Finally, thanks to you. There is so much choice out there,

so much to do and so many words to read that I count myself lucky you decided to read mine. Thank you. I'm glad you're here. Otherwise it'd just be me talking to the earwigs.